Morality, Foresig|

"Thinking about existential risk ... ... also fraught with risk itself. When it comes to technology developments, the ones with the greatest impact are usually the ones that are the most unanticipated. Nevertheless, as Louis Pasteur said, 'fortune favors the prepared mind,' and unless we try and prepare as carefully as we can for a future in which technology evolves at an exponential rate, the likelihood that the future could bring catastrophe on a global scale will increase. This book presents a sober and careful examination of the emerging field of existential risk studies, and will provide a useful introduction to all those who want to come up to speed quickly on developments over the past decade."

—**Lawrence M. Krauss**, Director of the Origins Project at Arizona State University, and Chair of the Board of Sponsors of the *Bulletin of the Atomic Scientists*. His most recent book is *The Greatest Story Ever Told... So Far: Why are we here?*

"*Morality, Foresight, and Human Flourishing* is an excellent introduction to a new and important area of research. I hope it will be widely read."

—**Peter Singer**, Ira W. DeCamp Professor of Bioethics at Princeton University and author of *Animal Liberation* and *The Most Good You Can Do*

"The path to our future is rife with threats to the very existence of humanity. How can we avoid creating technologies that destroy us, as well as other global catastrophes? We need a roadmap, and this is precisely what Torres provides in this carefully thought-out and useful book."

—**Susan Schneider**, Associate Professor at the University of Connecticut, author of *The Language of Thought: A New Philosophical Direction*, and editor of *Science Fiction and Philosophy*

"The exponential development of information technology promises extraordinary benefits for humanity, from the elimination of disease to radical life extension. But intertwined with this promise is great peril—existential risks associated with 'GNR' (genetics, nanotechnology, and robotics or AI). This book offers a careful exploration of this promise-versus-peril challenge. It is a must-read for anyone concerned about the future of humanity—and beyond."

—**Ray Kurzweil**, inventor, futurist, and author of *The Singularity Is Near* and *How to Create a Mind*

"Morality, Foresight, and Human Flourishing is an exceedingly thought-provoking book on a topic that must garner humanity's attention, namely the potential extinction of our species. This well-researched and well-written book is necessarily transdisciplinary because contributions pertinent to this topic entail perspectives from philosophy to climate change, and artificial intelligence to cognitive science. It is a must-read for those concerned with moving the world from myopic, crisis-driven policymaking to the proactive decision making needed to protect the future of humanity."

—**Bruce Tonn**, Consulting editor for the journal *Futures* and president of the nonprofit Three[3]

"Taking the reader on a harrowing yet hopeful tour of the landscape of existential risks, Torres masterfully molds complex, often abstract—but critically important—ideas about our continued existence into a concrete introduction to the topic. . . . Absolutely essential reading for anyone with the curiosity to learn of the dangers that lie in wait for humanity and the courage to believe that we can act to avoid them."

—**Gary Ackerman**, director of the Unconventional Weapons and Technology Division, National Consortium for the Study of Terrorism and Responses to Terrorism (START); and associate of the Global Catastrophic Risk Institute

"A careful study of modern-day emerging risks by one of the real emerging thinkers of our time. A must-read for anyone who cares about the future of the planet—in other words, a must-read for all of us."

—**Rachel Bronson**, Executive Director and Publisher of the *Bulletin of the Atomic Scientists*

"For millions of years, prior to the mid-1950s, humans lacked the ability to create or avoid a global catastrophe of any sort. Now we can do both—and for so many different scenarios that we now need a thorough review of threats and options for avoidance. We need this both because some of these scenarios interact and because considerable planning and prioritization is vital. This book by Phil Torres provides this essential resource with insights into both the high-level philosophical and the 'how-to' detailed levels. We must work hard to persist and thrive."

—**George Church**, Robert Winthrop Professor of Genetics at Harvard Medical School and advisor for the Future of Life Institute

"How might our world end? By our own hand? Or might danger loom from outside forces? Either way, Phil Torres takes you on a tour of the new field of Big Risk assessment, including ways we might protect our fragile promise."

—**David Brin**, physicist and award-winning science fiction author

"A primer for existential risk in the twenty-first century, including how humans are now a hazard to ourselves, not only as individuals but as a species. Read this book at your risk—and probably not before bedtime."

—**Jennifer Jacquet**, Assistant Professor of Environmental Studies at New York University and author of *Is Shame Necessary? New Uses for an Old Tool*

"*Morality, Foresight, and Human Flourishing* offers an authoritative guide to the emerging scientific discipline of existential risk in all its guises. . . . Highly recommended."

—**David Pearce**, moral philosopher and author of *The Hedonistic Imperative*

Morality, Foresight, and Human Flourishing

An Introduction to Existential Risks

Phil Torres

Foreword by Martin Rees

PITCHSTONE PUBLISHING
Durham, North Carolina

Pitchstone Publishing
Durham, North Carolina
www.pitchstonepublishing.com

10 9 8 7 6 5 4 3 2 1

Library of Congress Cataloging-in-Publication Data

Names: Torres, Phil, author.
Title: Morality, foresight, and human flourishing : an introduction to
    existential risks / Phil Torres ; foreword by Martin Rees.
Description: Durham, North Carolina : Pitchstone Publishing, [2017] |
    Includes bibliographical references and index.
Identifiers: LCCN 2017024484| ISBN 9781634311427 (pbk. : alk. paper) | ISBN
    9781634311441 (epdf) | ISBN 9781634311458 (mobi)
Subjects: LCSH: Risk—Sociological aspects. | Natural disasters. |
    Environmental disasters. | Technology—Risk assessment.
Classification: LCC HM1101 .T67 2017 | DDC 363.34—dc23
LC record available at https://lccn.loc.gov/2017024484

Cover design by Ian Thomas

Be curious and care.

Contents

Foreword

This is a welcome and timely book that draws attention to issues that our civilization's entire fate may depend on—and that need far more study and focus than they currently receive.

Our Earth is 45 million centuries old. But this century is the first when one species—ours—can determine the biosphere's fate. We're deep in a new era called the Anthropocene, where the main threats come not from nature, but from ourselves. In the crises of the Cold War era, the probability of stumbling toward Armageddon was put by some as high as one in three. That's tens of thousands of times higher than for an equally catastrophic asteroid impact.

Those of us with cushioned lives in the developed world fret too much about improbable air crashes, carcinogens in food, low radiation doses, and so forth. Current terrorism disproportionately fills the headlines. But we're in denial about far more shattering scenarios that thankfully haven't yet happened, but could.

The "x-risks" that threaten us are of two kinds. First, a growing population, more demanding of food, energy, and other natural resources, is putting unsustainable pressure on ecosystems, threatening loss of biodiversity and the crossing of climatic "tipping points."

But there's a second class of threats that will loom even larger: those stemming from the misuse, by error or design, of ever more powerful technologies. Nuclear weapons are based on twentieth-century science. But twenty-first-century sciences—biotech, cybertech, and artificial intelligence (AI)—will pose risks that are even more intractable.

Advances in genetics and microbiology offer exciting prospects, but they have downsides. It's accepted that techniques like "gain-of-function" modification of viruses and CRISPR/Cas9 gene editing will need regulation. There are precedents here: in the early days of recombinant DNA research, a group of biologists formulated the Asilomar Declaration, setting up guidelines on what experiments should and shouldn't be done. In the same spirit there's a call for similar regulation of the new techniques. However, the research community today, 40 years after Asilomar, is far larger, far more broadly international, and far more influenced by commercial pressures. Whatever regulations are imposed, on prudential or ethical grounds, could never be fully enforced worldwide—any more than the drug laws or tax laws can. Whatever can be done will be done by someone, somewhere. And that is deeply scary.

In consequence, maybe the most intractable challenges to all governments will stem from the rising empowerment of tech-savvy groups (or even individuals), by bio- as well as cybertechnology. This will aggravate the tension between freedom, privacy, and security.

These bio-concerns are relatively near-term—within 10 or 15 years. What about robotics and AI? Cyber threats are of course already pervasive and costly. And though we don't yet have the human-level robots that have been a staple of science fiction for decades, some experts think they will one day be real. If they could infiltrate the Internet—and the Internet of things—they could manipulate the rest of the world. They may have goals utterly orthogonal to human wishes—or even treat humans as an encumbrance. So how can we ensure that ever more sophisticated computers remain docile "idiot savants" and don't "go rogue"?

Experts disagree on how long it will take before machines achieve general-purpose human-level intelligence. Some say 25 years. Others say never. The median guess in a recent survey was about 50 years. And it's claimed that once a threshold is crossed, there will be an intelligence explosion. That's because electronics is a million times faster than the transmission of signals in the brain, and because computers can network and exchange information much faster than we can by speaking.

There is perhaps a parallel with nuclear fusion. Making an explosion—an H-bomb—has proven much easier than controlling it: the quest for controlled fusion power is still struggling. Likewise, containing an intelligence explosion might be harder than creating it.

In regard to all these speculations, we don't know where the boundary lies between what may happen and what will remain science fiction. But it's crucial that we explore this issue—one that I have previously addressed on numerous occasions. Environmental degradation, extreme climate change, or unintended consequences of bio-, cyber- and AI technology could trigger serious, even catastrophic, setbacks. We may have a bumpy ride through this century. We've no grounds for assuming that human-induced threats worse than those on our current risk register are improbable: they are newly emergent, so we have a limited time base for exposure to them and can't be sanguine about the ability to cope if disaster strikes. Moreover, in our interconnected world, the consequences would cascade globally.

It is crucial to focus more attention on these x-risks, and that is why this book is so timely. Phil Torres gives a comprehensive survey of the possible risks that have been discussed. He offers a clear (but scary!) review of the technologies. He also notes that the risk level depends on the number of humans who have the motivation to generate global terror—and, more mundanely, on the vulnerability of ever more complex systems to breakdown as well as innocent error.

There are already established research groups and government bodies addressing more "routine" risks—indeed, most organizations are required to produce a "risk register." But these extreme high consequence/low probability risks, potentially affecting the whole world, have hitherto been seriously addressed by only a small community of serious thinkers, whose ideas are described in the book. There needs to be a much expanded research program, involving natural and social scientists, to compile a more complete register of possible "x-risks," to firm up where the boundary lies between realistic scenarios and pure science fiction, and to enhance resilience against the more credible ones. The stakes are so high that those involved in this effort will have earned their keep even if they reduce the probability of a catastrophe by a tiny fraction.

Technology brings with it great hopes but also great fears. We mustn't forget an important maxim: the unfamiliar is not the same as the improbable.

This encyclopedic book is especially needed. Let's hope it has a wide resonance—and encourages a more intensive and serious focus on issues on which, it's no exaggeration to say, the fate of future generations depends.

—**Lord Martin Rees**, Astronomer Royal, former president of the Royal Society, member of the Board of Sponsors of the *Bulletin of the Atomic Scientists*, and cofounder of the Centre for the Study of Existential Risk

Preface

The field of existential risk studies can trace its origins back to the end of World War II, when the *Bulletin of the Atomic Scientists* created the Doomsday Clock to represent our collective nearness to a global disaster. Later, the astrobiologist Carl Sagan popularized the Drake equation (section 1.5) in the television series *Cosmos* and published an important commentary on the consequences of a major nuclear conflict.[1] According to Sagan, if humanity survives for the next 10 million years, we could expect some 500 trillion people to come into existence.[2] Thus, an all-out nuclear exchange that causes human extinction would not only kill the entire current human population but close off the possibility of *billions and billions* of future lives ever being lived. This makes extinction scenarios especially worrisome—a class of catastrophes with unique moral significance.[3]

In the mid-1990s, the Canadian philosopher John Leslie published an important book called *The End of the World: The Science and Ethics of Human Extinction*, which covers a wide range of existential risks—although he didn't use that term. Leslie also provided perhaps the most compelling defense to date of the doomsday argument (section 7.1), which implies that we are systematically underestimating the probability of human extinction. The work of Leslie influenced another notable figure, namely, Nick Bostrom, the founding director of the Future of Humanity Institute (FHI) at the University of Oxford. Bostrom's work initially focused on anthropic reasoning, including the observation selection effect (section 1.6), which has some important implications for evaluating the overall risk of annihilation.

In 2002, Bostrom published an article in the *Journal of Evolution and Technology* called "Existential Risks: Analyzing Human Extinction Scenarios and Related Hazards." This formalized the concept of an existential risk, introduced the Maxipok rule (section 1.4), and offered an authoritative outline of the biggest threats to our collective future. Bostrom's 2002 article is largely responsible for the popularity—and publicity—of existential risk studies today, a feat that was helped along by his 2014 best seller *Superintelligence*, which provides a detailed account of the technical and philosophical challenges of creating a "friendly" superintelligence.

Although one could argue that the field hasn't quite reached a "normal science" mode of operation yet—to borrow a term of art from Thomas Kuhn—there is an emerging consensus about the central terms, fundamental concepts, and canonical works of existential risk scholarship.[4] There has also been an explosion of institutes dedicated to (a) studying the various existential risks that haunt our species, and (b) devising strategies to mitigate these risks. Such research organizations include the aforementioned FHI as well as the Centre for the Study of Existential Risk (CSER), Future of Life Institute (FLI), Global Catastrophic Risk Institute (GCRI), and my own X-Risks Institute (XRI). In some cases, high-profile scholars or celebrities have put their weight behind these organizations to increase public awareness. For example, Stephen Hawking, Alan Alda, and Morgan Freeman are all members of FLI's scientific advisory board.

So, the "x-risk ecosystem," as the cofounder of FLI and CSER Jaan Tallinn calls it, has grown into a thriving network of scholars and institutions bridging both popular culture and academia.[5] Yet the field does not so far have a comprehensive "textbook" to guide curious young scholars who would like to make the greatest possible impact on the world.[6] This book—an advanced introduction to existential risks; essentially, a progress report on the field—aims to fill this lacuna, thereby further establishing the field as a legitimate area of intellectual inquiry. It attempts to adumbrate something *resembling* a "paradigm" by integrating a wide range of ideas that bear on the topic. (See the postscript for discussion.)

The target audience includes undergraduate and graduate students in fields as diverse as philosophy and ethics, political science, engineering, computer science, cognitive science, psychology, terrorism studies, sociology, cosmology, and risk analysis.[7] In addition, policymakers, politicians, entrepreneurs, and other culture shapers should find this book full of timely and useful insight.[8] More than anything, I would like *Morality, Foresight, and Human Flourishing* to inspire bright minds around the globe to think more, and more carefully, about the possible, probable, and preferable futures of our species on this planet—and beyond.[9]

Chapter 1: An Emerging Field

1.1 A Unique Moment in History

One can make a very strong case that humanity has never lived in more peaceful times. According to the Harvard polymath Steven Pinker, violence has been declining since humanity struggled as hunter-gatherers in the Paleolithic, roughly 12,000 years ago. This trend has continued through the twentieth and into the twenty-first century, despite the two world wars, Korean War, Vietnam War, Second Congo War (also known as the African World War), and rise of global terrorism, associated most notably with al-Qaeda, Boko Haram, and the Islamic State. We find ourselves in the midst of (a) what historians call the "Long Peace," a period that began at the end of World War II and during which no two superpowers have gone to war, and (b) what Pinker tentatively dubs the "New Peace," which refers to "organized conflicts of all kinds—civil wars, genocides, repression by autocratic governments, and terrorist attacks—[having] declined throughout the world" since the Cold War concluded in 1989.[1] If you could choose when you would like to live in human history since our debut in East Africa some 200,000 years ago, the most reasonable answer would be, "Today, at the dawn of the twenty-first century. No question!"[2]

But there is a countervailing trend that tempers the good news presented by Pinker's historical analyses: we might also live in the most dangerous period of human history, ever.[3] The fact is that our species is haunted by a *growing swarm of risks* that could either trip us into the eternal grave of extinction or irreversibly catapult us back into the Stone Age. Just consider that humanity has stood in the flickering shadows of a nuclear holocaust since 1945, when the United States dropped two nuclear bombs on the Japanese archipelago. In the years since this epoch-defining event, scientists have confirmed that climate change and global biodiversity loss are urgent threats with existential implications, while risk experts have become increasingly worried about the possibility of malicious individuals creating designer pathogens that could initiate a worldwide pandemic. Looking further along the threat horizon, there appears to be a number of unprecedented dangers associated with molecular nanotechnology and artificial intelligence.[4] Thus, one only needs simple arithmetic to see that the total number of *existential risk scenarios* has increased significantly since the Atomic Age began, and it looks as if this trend will continue at least into the coming decades, if not further.[5]

Considerations of these phenomena have led some scholars to offer unsettlingly high estimates that a global disaster will occur in the foreseeable future.[6] For example, the philosopher John Leslie argues that we have a 30 percent chance of extinction in the next five centuries.[7] Even more ominously, an "informal" 2008 survey of experts at a conference hosted by the Future of Humanity Institute gave a 19 percent chance of extinction before 2100.[8] And the cosmologist Martin Rees writes in a 2003 book that civilization has a 50-50 chance of surviving the present century.[9] To put this in perspective, consider that the average American has a 1-in-9,737 lifetime chance of dying in an "air and space transport accident."[10] It follows that according to the FHI survey, the average American is at least *1,500 times more likely* to perish in a human extinction catastrophe than a plane crash. Using Rees's estimate, the average American is nearly *4,000 times more likely* to encounter a civilizational collapse than to die in an aviation mishap.[11]

If this sounds unbelievable—and no doubt it does, and should—reflect on how many people would be affected by such a disaster. An analogous case involves asteroids (see section 2.4). According to statisticians, the average person is more likely to die from an asteroid impact than a bolt of lightning (which itself is more likely to kill the average American than a terrorist attack). In fact, the U.S. National Research Council reports that we should *expect* an average of 91 deaths each year from asteroids striking Earth, even though the *actual* number is almost always zero.[12] They calculate this number by considering how many asteroids there are near Earth, how big these asteroids are, and how devastating an impact would be. Averaging the total expected deaths over millennia, they get the counterintuitive results above.[13] So, the comparisons of the previous paragraph might not be that far off the mark: a child born today may have a very good chance of living to see global society destroy itself.[14]

Finally, consider the Doomsday Clock, a metaphor that represents our collective nearness to doom, or midnight. This clock was created in 1947 by the *Bulletin of the Atomic Scientists*, an organization founded by physicists who had previously worked on the Manhattan Project, which built the first atomic bombs. Over time, the minute hand of the clock has moved back-and-forth to track the vicissitudes of world affairs: beginning at 7 minutes to midnight in 1947, it moved to only 2 minutes in 1953 (after the United States and Soviet Union both detonated hydrogen bombs) and then drifted away from doom to 17 minutes before midnight when the Cold War "officially" ended in 1991.[15]

While the Bulletin was originally founded to monitor the dangers posed by the world's nuclear arsenals, it announced in 2007 that "climate change also presents a dire challenge to humanity." Consequently, the clock's minute hand inched from 7 to 5 minutes to midnight. After wavering between 5 and 6 minutes, it moved forward again in 2015 due to "unchecked climate change, global nuclear weapons modernizations, and outsized nuclear weapons arsenals," which "pose extraordinary and undeniable threats to the continued existence of humanity." A year later, the Bulletin decided to keep the clock set at 3 minutes to midnight, writing that "the world situation remains highly threatening to humanity, and decisive action to reduce the danger posed by nuclear weapons and climate change is urgently required."[16] But 2017 saw the minute hand tick 30 seconds closer to doom, reaching the highest level of danger since 1953. This was largely due to two factors, both enabled by what one could describe as a *zeitgeist of anti-intellectualism* that currently pervades Western, especially American, political culture. As the Bulletin's official statement puts it, an *already-threatening world situation was the backdrop for a rise in strident nationalism worldwide in 2016, including in a U.S. presidential campaign during which the eventual victor, Donald Trump, made disturbing comments about the use and proliferation of nuclear weapons and expressed disbelief in the overwhelming scientific consensus on climate change.*[17]

On the same day of this announcement, the cosmologist Lawrence Krauss and international affairs expert David Titley, both of whom help maintain the Doomsday Clock, published a *New York Times* op-ed titled "Thanks to Trump, the Doomsday Clock Advances toward Midnight." In their words,

The United States now has a president who has promised to impede progress on both [curbing nuclear proliferation and solving climate change]. Never before has the Bulletin decided to advance the clock largely because of the statements of a single person. But when that person is the new president of the United States, his words matter.[18]

The point is that *many* leading experts believe the threat of an existential catastrophe to be significant.[19] Before 1945, overseeing a Doomsday Clock would have been utterly nonsensical, since the existential threats posed by nature are relatively improbable (see below). Yet today, the clock stands at two-and-a-half minutes before midnight, and it appears poised to tick forward again in 2018. To be sure, the predicament of *Homo sapiens* on Earth has always been precarious—consider that we are the *only remaining species* of *Homo* on the planet, our relatives the Neanderthals having died out about 40,000 years ago—but changes to the global climate and ecosystem along with the development of powerful new technologies are making our continued survival more uncertain than ever.

1.2 What Are Existential Risks?

The concept of an **existential risk** (ER) was formalized by the Oxford philosopher Nick Bostrom in a 2002 paper.[20] To understand this term's definition, it is helpful to know that Bostrom is a prominent figure within the *transhumanist movement*. According to transhumanism, person-engineering technologies will enable us, if we wish, to modify aspects of our bodies and brains, perhaps resulting in a new species of *posthumans*, while world-engineering technologies will enable us to radically redesign the environments in which we live to make them more conducive to flourishing (where some of these environments could be simulated rather than "real").[21] Whereas bioconservatives embrace "therapeutic" but not "enhancive" interventions on the human organism, transhumanists advocate exploring what could be a vast space of posthuman modes of being, some of which may be *far better* in certain moral respects than our current human mode.[22] Thus, transhumanism has both descriptive and normative components.[23] (See Box 1.)

To be clear, most transhumanists are careful to emphasize that "can" does not imply "ought"—that is, just because we are able to modify our phenotypes doesn't mean that we are obliged to do so. Rather, humanity should proceed according to something like the "precautionary principle," which states that "an action should not be taken if the consequences are uncertain and potentially dangerous,"[24] or perhaps the philosopher Max More's "proactionary principle," which argues that

> *People's freedom to innovate technologically is highly valuable, even critical, to humanity. This implies several imperatives when restrictive measures are proposed: Assess risks and opportunities according to available science, not popular perception. Account for both the costs of the restrictions themselves, and those of opportunities foregone. Favor measures that are proportionate to the probability and magnitude of impacts, and that have a high expectation value. Protect people's freedom to experiment, innovate, and progress.*[25]

Box 1. As the AI entrepreneur Riva-Melissa Tez puts it, transhumanism "sounds weirder than it actually is."* It is simply the idea that, within certain ethical boundaries and guided by the epistemic value of "philosophical fallibilism," we should not be afraid to use technology to improve the human condition, which is currently marked by widespread suffering, the hedonic treadmill, disease, senescence, and death. There are a couple of points worth noting here: First, we have already vastly improved our situation through the use of technologies, some of which—such as clothes, glasses, telescopes, prosthetics, psychoactive pharmaceuticals, pacemakers, cochlear implants, smartphones, and the Internet—directly alter, extend, and enhance our phenotypes. Compared to our Paleolithic progenitors, most modern humans are "transhumans" already—virtually a different species. Second, humanity is evolving anyway due to ongoing mechanisms like natural selection and genetic drift, and indeed some scientists believe that human evolution has actually accelerated in recent centuries. Thus, we will someday become "posthumans" even if bioconservative policies are universally implemented, just as some of our ancient Hominini relatives became "post-Australopithecines" by evolving into *Homo sapiens*. Since biological evolution is a non-teleological process—meaning that *every state* is an in-between state; there is no finalistic "resting place" at which all human genetic changes cease†—why not try to take control of our own evolution through intentional cyborgization, to direct our lineage toward future states marked by improved health, happiness, longevity, intelligence, morality, and so on? This isn't such a radical idea after all—and in fact one could argue that it is the default, albeit tacit, view of many Westerners today. It is certainly the direction in which our technological civilization appears to be headed.

*ogilvydo.2015.Technology:MakingtheWorldaBetterPlace.YouTube. URL: https://www.youtube.com/watch?v=i5t1BQUbSB4&t=43s.

† Or, as Charles Darwin put it, "not one living species will transmit its unaltered likeness to a distant futurity." Thus, there is a sense in which bioconservatism is a *nonstarter*. See Darwin, Charles. 2007. *On the Origin of Species: By Means of Natural Selection or The Preservation of Favored Races in the Struggle for Life.* New York, NY: Cosimo Classics.

Having outlined the basics of transhumanism, we can now make sense of Bostrom's definition of an existential risk:

> *An existential risk is one that threatens the premature extinction of Earth-originating intelligent life or the permanent and drastic destruction of its potential for desirable future development.*[26]

Thus, there are two general categories of existential risk scenarios: (i) total annihilation, and (ii) an irreversible curtailing of our potential. The first disjunct is straightforward: the lineage of Earth-originating intelligent life terminates. This outcome is *binary*: we either live or die, persist or desist, remain extant or go extinct. The second disjunct is not so clear-cut, given the normativity of "desirable." It is here that transhumanism enters the axiological picture. From this perspective, the ultimate goal of civilization is to safely reach a state of *technological maturity*, which Bostrom limns as "the attainment of capabilities affording a level of economic productivity and control over nature close to the maximum that could feasibly be achieved."[27] It follows that a catastrophe—in this case, an endurable catastrophe of type (ii)—counts as "existential" if and only if it prevents our species from realizing the posthuman promise of "mature technology."

In addition to a definition of "existential risk," Bostrom offers three typologies of risks in general.[28] These are based on a conceptual decomposition according to which a risk equals *the probability of an event multiplied by its consequences.* (Note that this entails that a high-consequence risk could be significant even if it is extremely improbable.) With respect to the first variable, there are multiple interpretations of probability, such as the propensity, frequency, and Bayesian interpretations, none of which we will here explore. With respect to the second, Bostrom analyzes the consequences of an event into two subcomponents: *scope* and *intensity*. Scope refers to how many people are affected, and intensity to how bad the effects are. The result is a two-dimensional typology, Figure A, in which existential risks occupy the top right box of transgenerational-terminal events (where "terminal" is stipulated to include some endurable events).[29]

Figure A. Two-Dimensional Typology of Risks

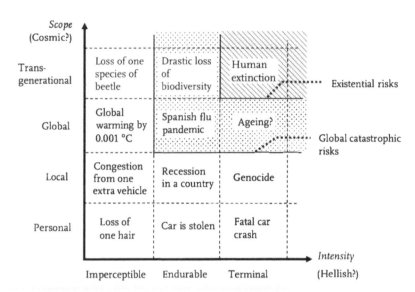

Source: Nick Bostrom and Milan Ćirković. 2008. Introduction. In Nick Bostrom and Milan Ćirković (editors), *Global Catastrophic Risks*. Oxford: Oxford University Press.

But we can refine Bostrom and (his coauthor) Milan Ćirković's typology by further decomposing the scope of a risk's consequences into *spatial* and *temporal* sub-subcomponents. This is motivated by the truism that risks can have a range of different spatiotemporal ramifications. For example, a germline mutation could have limited consequences within a population, yet these consequences could linger for an indefinite number of future generations. (Where would this risk fit in Figure A?) Similarly, a catastrophe could *instantaneously* kill 1 billion people at a given timeslice or *incrementally* kill the same number over the course of a century. Distinguishing between these scenarios is important because our responses to each might require quite divergent counterstrategies. Thus, insofar as Bostrom and Ćirković's typology is intended to provide an exhaustive classification of risks, it appears inadequate.[30]

By adopting a decomposition of risks according to the three properties of intensity, spatial scope, and temporal scope, one gets

Figure B. Alternative Typology of Risks Based on the Properties of Intensity (x-axis), Spatial Scope (y-axis), and Temporal Scope (z-axis)

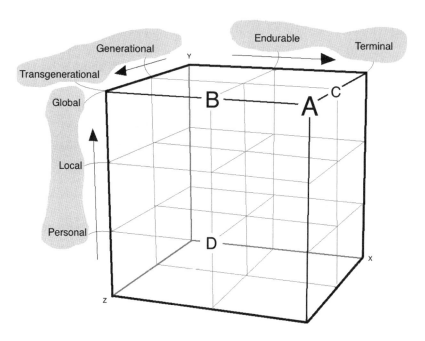

Note: Consequences get worse as one follows the arrows.

the three-dimensional typology of Figure B. In this figure, existential risks occupy two positions: (1) the node marked "A," which corresponds to global-*terminal*-transgenerational catastrophes, and (2) the node marked "B," which encompasses those global-*endurable*-transgenerational events that, by definition, prevent humanity from ever attaining technological maturity. Furthermore, germline mutations correspond to node D, while aging (which fits uncomfortably in Figure A, as indicated by the question mark) corresponds to node C—that is, it affects everyone globally with death but doesn't entail our extinction.[31]

Whichever typology one finds most useful, the key idea is that existential risks constitute *worst-case scenarios* for humanity—resulting in what the philosophers Ingmar Persson and Julian Savulescu call "Ultimate Harm"—given our potential to reach new and better modes of being.[32]

Other important features of existential risks are the following:

(a) They are *singular events* that can only happen once in a species' lifetime; this makes them quite unique among all the other types of risks that we face. For example, we can talk about a human extinction event happening tomorrow but not about it having happened yesterday; and while we can talk about an endurable existential catastrophe having happened yesterday, we would not be able to do anything about it, to reverse the outcome. If an existential risk were to occur, *the game would be over and humanity would have lost.*

(b) Since existential risks have the properties of (a), our strategies for avoiding them must rely entirely on *anticipation* rather than *retrospection.* As Bostrom writes, "The reactive approach—see what happens, limit damages, and learn from experience—is unworkable. Rather, we must take a proactive approach."[33] This means that humanity must employ "unnatural" modes of thinking, since our typical way of avoiding bad future circumstances is to update our world models in response to past mistakes made by ourselves or others. But there is no possibility of learning from the mistakes that humanity made leading up to an existential catastrophe so that we don't encounter another existential catastrophe later on.[34]

(c) These points suggest that individuals and governments are unlikely to make existential risk reduction a top priority. Since an effective risk mitigation program would result in the *absence* rather than *presence* of an observable event, a record of success could lead to complacency, causing people to question whether money is being well-spent. The risk analyst Nassim Taleb makes this point in the context of "black swans," or game-changing incidents that are inadequately expected:

> It is difficult to motivate people in the prevention of Black Swans. . . . Prevention is not easily perceived, measured,

or rewarded; it is generally a silent and thankless activity. Just consider that a costly measure is taken to stave off such an event. One can easily compute the costs while the results are hard to determine. How can one tell its effectiveness, whether the measure was successful or if it just coincided with no particular accident? . . . Job performance assessments in these matters are not just tricky, but may be biased in favor of the observed "acts of heroism." History books do not account for heroic preventive measures.[35]

(d) Even more, the reduction of existential risks constitutes a *global public good*, meaning that it is both *non-excludable* (i.e., it is not possible to prevent those who haven't paid for this service from benefiting) and *non-rivalrous* (i.e., it is not the case that one person benefiting prevents others from benefiting). This is notable because markets don't typically provide such goods, since producers can only retrieve a small amount of value relative to the costs of production. As Bostrom elaborates this point,

> *In fact, the situation is worse than is the case with many other global public goods in that existential risk reduction is a strongly* transgenerational . . . *public good: even a world state may capture only a small fraction of the benefits—those accruing to currently existing people. The quadrillions of happy people who may come to exist in the future if we avoid existential catastrophe would be willing to pay the present generation astronomical sums in return for a slight increase in our efforts to preserve humanity's future, but the mutually beneficial trade is unfortunately prevented by the obvious transaction difficulties.*[36]

So, existential risks form a special class of catastrophes that pose genuinely unique challenges to civilization.

Before moving on to the next section, we should consider a related topic of interest, namely, **global catastrophic risks** (GCRs). Bostrom and Ćirković define GCRs "loosely" as events "that might have the potential to inflict serious damage to human well-being on a global scale." They suggest that a disaster causing "10 million fatalities or 10 trillion dollars worth of economic loss . . . would count as a global catastrophe, even if some region of the world escaped unscathed."[37] Other scholars have defined GCRs as events that result in one-fourth of the human population dying, or "threats that can eliminate at least 10% of the global population."[38] In Figure A, GCRs encompass risks within the light and dark gray boxes—meaning that existential risks are a special case of global catastrophic risk. With respect to Figure B, we can define GCRs as any risk that (a) has the property of *being global*—that is, it instantiates a node on the top level of the diagram—and (b) causes sufficiently severe harm to human civilization.[39]

Given that the probability of a risk tends to *increase* as its consequences *decrease*, the chance that one or more GCRs occur this century should exceed the probabilities assigned to an existential catastrophe occurring—which, once again, range from 19 percent to 50 percent.[40] More concretely, a pandemic that kills 1 billion people will be more probable than one that causes human extinction; the same goes for an asteroid impact, nuclear war, nanotech accident, and so on. Thus, if we believe that human extinction from a pandemic has, say, a 1 percent chance of happening per decade, we should believe that 1 billion people dying in a pandemic has a *greater than or equal to 1 percent chance* of happening over the same period.[41] In general, the smaller the consequences, the higher the probability.

Furthermore, insofar as the *timing* of non-existential GCRs is random—which is not an implausible assumption, since (a) many natural risks are in some sense "random," and (b) studies have actually shown that "the onsets of wars [are] randomly timed"—we should (weakly) expect them to cluster together in time.[42] For example, if there is a constant probability of 0.05 that a GCR will occur per decade, and if a GCR occurs during the first decade of a new century, the probability of a GCR occurring the second decade will actually be *higher* than one occurring the third decade, or any decade afterwards. The reason is that for a GCR to occur next during the third decade, it would have to *not have occurred* during the second. Thus, *two conditions* must hold: (i) no GCRs during the second decade, and (ii) a GCR during the third decade. To calculate the probability of this joint state of affairs, one multiplies the probability of (i), or 0.95 (from 1 minus 0.05), by the probability of (ii), or 0.05. This yields a probability of 0.0475 for a GCR happening in the third decade, which is, of course, lower than the 0.05 probability of a GCR happening in the second decade. As the mathematician William Feller once put it, "To the untrained eye, randomness appears as regularity or tendency to cluster," meaning that we should not be *too* surprised if a series of global catastrophes unfolds one after another.[43]

While this book focuses primarily on existential risks, given their unique moral status (see section 1.4), GCR issues will nonetheless appear throughout.

1.3 Types of Existential Risks

There are different ways to taxonomize existential risks depending upon one's theoretical or practical goals. In a 2013 paper, Bostrom offers a four-part scheme that includes human extinction, permanent stagnation, flawed realization, and subsequent ruination. With respect to Figure B, the first is a global-terminal-transgenerational disaster (node A), whereas the latter three are global-endurable-transgenerational disasters (node B). Taking these in turn:

(i) **Human extinction**. About 99.9 percent of all species that have ever existed on Earth have gone extinct, and the average mammal survives for only about 2.5 million years.[44] As Carl Sagan put it, "Extinction is the rule. Survival is the exception."[45] Here we should expand the semantics of "human" to include not just *Homo sapiens* but Earth-originating intelligent life in general, independent of its material substrate (e.g., living cells or microchips). This is important because if the cyborgization trend of integrating biology and technology, organism and artifact, continues, our descendants could become sufficiently different from us to constitute a new species: *Homo cyborgensis*, or something of the sort.[46] If a future posthuman population of *Homo cyborgensis* were completely decimated, we should like this to count as an existential catastrophe too.

(ii) **Permanent stagnation.** This scenario would occur if (i) does not obtain yet humanity never reaches a state of technological maturity. Bostrom distinguishes several types of stagnation, including (a) *unrecovered collapse*, where "much of our current economic and technological capabilities are lost and never recovered," (b) *plateauing*, where "progress flattens out at a level perhaps somewhat higher than the present level but far below technological maturity," and (c) *recurrent collapse*, which would entail "a never-ending cycle of collapse followed by recovery."[47] To this taxonomy we can add a "catch-all" category that includes any combination of these scenarios, such as long plateaus punctuated by collapse, followed by recovery to another plateau, followed by unrecovered collapse.

(iii) **Flawed realization.** This involves reaching "technological maturity in a way that is dismally and irremediably flawed." In other words, we achieve a posthuman state that realizes only "a small part of the value that could otherwise have been realized."[48] Bostrom identifies two instances of this outcome. The first, *unconsummated realization*, occurs when future technologies fail to achieve states of high value. For example, it could be the case that future artificial intelligences (AIs) inherit the world, but that these AIs do not have conscious experiences like we do. As the philosopher Susan Schneider rightly emphasizes, a world full of unconscious machines—even if these machines were to build a complex, advanced civilization throughout the known universe—would be far less valuable than one in which even a single conscious being exists.[49] The result would be an existential catastrophe.

The second type of flawed realization is *ephemeral realization*. This results when "humanity develops mature technology that is initially put to good use. But the technological maturity is attained in such a way that the initial excellent state is unsustainable and is doomed to degenerate." For example, it could be that achieving technological maturity leads to significant social, political, or cultural divisions that over time cause major conflicts

Box 2. Of all the existential risk categories here enumerated, extinction appears to be the least likely. The reason pertains to what might be called the *last few people problem*: one can readily devise hypothetical narratives in which a large number of humans perish, but it is rather hard to envision how the last people on the planet follow their conspecifics to the grave. This problem emerged from a 2009 special issue of the journal Futures, co-edited by Bruce Tonn and Donald MacGregor, in which scholars were tasked with concocting extinction scenarios. As Tonn and MacGregor write, "It is quite easy to imagine events that could lead to a rapid and massive loss of human life. . . . [But most] of the scenario writers found that indeed it was difficult to kill off the last few humans and most were surprised . . . for this to be the case. We speculate that is the good news coming out of this special issue."*

* Tonn, Bruce, and Donald MacGregor. 2009. Are We Doomed? *Futures*. 41(10): 673–675.

to break out, and that these conflicts bring about an extinction or permanent stagnation disaster. As Bostrom puts it, "There is a flash of value, followed by perpetual dusk or darkness."[50]

(iv) **Subsequent ruination.** Our final category occurs when (i) through (iii) fail to obtain, meaning that we reach an unflawed state of technological maturity. Our species appears to have accomplished the ultimate triumph. Yet *further developments* in technology, social institutions, government, and so on bring about either the termination of our lineage or an irreversible decline in our quality of life.[51] (See Box 2.)

While this taxonomy is helpful for understanding different features of possible worst-case futures, we will adopt a different approach that focuses not on the *outcomes* of various scenarios but on those scenarios' *causes*. We can call this the **etiological approach**. Attending to the underlying causes of different scenarios is arguably more important because when one understands the causes behind an effect, one can avoid the effect by intervening on the causes. For example, if you know that a brake failure was the cause of your car racing through a red light, then you can prevent future traffic violations by fixing the brakes. Similarly, if you know that smoking causes lung cancer, then you can reduce your chances of a bad oncological diagnosis by refraining from smoking. Thus, specifying the etiology of different outcomes is crucial for avoiding a catastrophe.

The broadest causal distinction is between **natural risks** and **anthropogenic risks**. Supervolcanic eruptions, natural pandemics, and asteroid or comet impacts are the most worrisome natural risks. Less concerning are supernovae, gamma-ray bursts, galactic center outbursts, superstrong solar flares, and black hole mergers or explosions. The universe could also contain any number of currently unknown risks to our survival. Perhaps a discovery by physicists 50 years from now will reveal a new type of natural danger that is as unimaginable to twenty-first-century humans as the threat of gamma-ray bursts was to those in the Pleistocene. Or it could be that no possible future science can reveal certain threats because understanding them requires a different *kind of mind* than what natural selection gave us. As far as contemporary science is concerned, though, the overall probability of a natural existential risk destroying humanity per century is almost certainly less than 1 percent, and arguably *far less*.[52]

Moving on to the category of anthropogenic risks, this contains a diverse range of distinct and overlapping phenomena. The most significant subtype stems from what we will refer to as **agent-tool couplings**.[53] We can define an agent somewhat crudely as any entity capable of making its own decisions in pursuance of its own goals, whatever they happen to be. There are many *degrees* of agency in the world: for example, a heat-seeking missile has a certain degree of agency since it can navigate space-time in response to inputs relating to its target. The agents most relevant in this context, though, are those with general intellectual abilities, whether human or machinic in nature, such as apocalyptic terrorists and artificial superintelligence. As for the tool half of the coupling, this includes any advanced technology with the capacity to cause an existential catastrophe. We can call these **weapons of total destruction** (WTDs), on the model of "weapons of mass destruction" (WMDs). Such technologies could be actual (e.g., nuclear weapons) or merely anticipated (e.g., molecular nanotechnology), and indeed many existential risk scholars believe that future anticipated technologies will likely pose far greater risks than those around today. There could also be technologies that are not currently anticipated by anyone but that will introduce novel hazards for humanity.

The "agent-tool" concept is essential for existential risk studies because, bracketing the possibility of malfunction, dangerous technologies require a suitable agent to *use* them to cause harm. It follows that to assess the relevant risks, one must evaluate *both* the artifacts *and* their users. This framework also emphasizes that there are two definite variables—the agents and the tools—that could be intervened upon to reduce overall existential risk. Thus, chapter 6 explores "tool-oriented" and "agent-oriented" strategies for reducing existential risks.

Another subtype of anthropogenic risk derives from **unintended consequences**. The most troubling unintended consequences today are climate change and biodiversity loss, although there are also potential risks associated with physics experiments and geoengineering. As all responsible citizens of the world should know, climate change is the result of greenhouse gas emissions, which are a byproduct of burning fossil fuels. This is arguably the first unintended consequence in human history with genuinely existential implications—but it will probably not be the last. Indeed, when automobiles with internal combustion engines were adopted *en masse* in the early twentieth century, they were widely praised as a solution for urban pollution, a major health problem at the time, which took the form of horse excrement and carcasses. (This also resulted in the spread of illness by the "disease vectors" of flies.[54]) The unfortunate irony is that automobiles have become one of the greatest contributors to a global-scale calamity that threatens the future stability of civilization itself. While climate change is a primary cause of biodiversity loss, which has initiated a new mass extinction, biodiversity loss can also exacerbate climate change—for example, through the elimination of carbon sinks, which remove carbon dioxide from the atmosphere.

As for physics disasters: while this scenario appears highly improbable on our best current theories, these theories could be flawed. A high-powered particle accelerator could thus accidentally initiate a catastrophe with planetary or even *cosmic* consequences. Geoengineering, which involves redesigning one or more physical features of our planetary spaceship (i.e., Earth), poses several perils. For instance, a group or government could unilaterally opt to inject particles into the stratosphere to block incoming sunlight, thereby reducing the negative consequences of "too much" atmospheric carbon dioxide. Although this could, it appears, save humanity in a climate emergency (see section 6.5), it could also have severe unintended repercussions. Alternatively, if the injection of particles into the stratosphere were to work but then suddenly stop for some reason, surface temperatures could rebound too quickly for civilization to adapt.

Finally, we will examine a range of risks that don't directly arise from either agent-tool couplings or the unintended consequences of human activity. This motley group includes:

(a) *Simulation shutdown.* However dubious this may initially sound— and it should sound dubious to any good skeptic—there are some rather compelling, albeit esoteric, reasons for believing that we might live in a computer simulation. If so, this would introduce the possibility that our simulation gets shut down, thereby resulting in an existential catastrophe.[55]

(b) *Bad governance.* Unwise governments could ignore the established science behind climate change and biodiversity loss—and, indeed, many governments are doing precisely this. They could also engage in arms races involving molecular nanotechnology or superintelligence, both of which would likely yield "winner-take-all" situations. If such a race were to occur and if the "winner" were to "take all," humanity could find itself under the control of a totalitarian state—one that might stifle further technological development, not to mention human happiness.

(c) *Something completely unforeseen.* It would be imprudent to believe that we—apes with big foreheads—know all the risks to our species. There could be unknown natural risks, unanticipated future technologies, new types of dangerous agents, and unintended consequences from, say, colonizing space. A book like this written 200 years from now could contain 3 (or 20) times as many chapters focusing on scenarios of which we haven't the slightest inkling. Indeed, the existential risks explored throughout this manuscript could be relegated to the appendix, being seen as the *least worrisome* relative to the new, futuristic threat environment of our descendants.

* * *

There are a few conceptual distinctions worth mentioning before moving on. First, consider the difference between **state risks** and **step risks**.[56] The former arise from being in a certain state, whereas the latter arise from transitioning between states. To illustrate, dying in a car accident is a state risk: the danger is associated with a specific situation, namely, driving a car, and the longer that one is in this situation, the greater the risk. Many risks from nature are state risks. In contrast, walking onto a train from the railway platform constitutes a step risk: the danger is associated with the transition from being on the platform to being in the train. Thus, in the London Underground one hears the warning, "Mind the gap." Once inside the train, the danger is gone (although one then encounters a new state risk). The existential danger posed by superintelligence may be a step risk.

There are also what we might call **context risks**. These are big-picture phenomena that *frame* our existential predicament on the planet. The most notable context risks are climate change and biodiversity loss. Such risks have implications for the overall probability of doom, even if they are themselves unlikely to bring about an existential catastrophe (that is, as a proximate cause of the disaster). Put differently, contexts risks can *modulate* the dangers posed by other risk scenarios. A simple intuition pump illustration is the following: imagine two worlds, A and B. World A finds itself beset by social turmoil, economic meltdowns, and political strife as a result of environmental atrophy, whereas the climate and biosphere of world B remain in relative homeostasis. Now imagine that both worlds contain 10,000 nuclear weapons. In which world is nuclear conflict more likely to break out *a priori*? The answer is, obviously, world A. The capacity for conflict-multiplying context risks to raise or lower the tide of all other existential threats makes phenomena of this sort especially important to prioritize. (This is a crucial point that I hope readers will dwell on.)

1.4 Why Care about Existential Risks?

> *Nothing is too wonderful to be true,*
> *if it be consistent with the laws of nature.*
> —*Michael Faraday*

The global population today is 7.5 billion. Let's say that a pandemic spreads across Europe, killing 100 million people. How bad would this be? Most would agree that it would be quite devastating. Now let's say that 100 million more people die from the disease. How bad would *this* be? It seems like this second wave of deaths would be just as bad as the first: 200 million people dying is twice as horrible as 100 million people dying. Now imagine this continuing 74 times (where 74 x 100 million = 7.4 billion), with each instance of 100 million deaths being an equivalently bad moral tragedy. The global population would then be only 100 million people. Again, we can ask: If this last group were to die from the pandemic, how bad would it be? Would it be just as bad as each past instance of 100 million people dying—or might it be *worse*?

The philosopher Derek Parfit, echoing Sagan's idea discussed in the preface of this book, argues that the last 100 million people dying would not only be worse than all the other instances of 100 million people dying, but *profoundly worse*. The reason is that, as Parfit writes,

> *Civilization began only a few thousand years ago. If we do not destroy mankind, these few thousand years may be only a tiny fraction of the whole of civilized human history. The difference between [nearly all and actually all people dying] may thus be the difference between this tiny fraction and all of the rest of*

this history. If we compare this possible history to a day, what has occurred so far is only a fraction of a second.[57]

We can add to Parfit's thesis an alternative scenario, given the second disjunct of our definition of existential risks: consider a world in which there are no incidents of mass dying but some unfortunate event causes civilization to sink into a permanent state of technological deprivation. The result would be that we fail to reach technological maturity and exploit our *cosmic endowment of negentropy* (where "negentropy" is a portmanteau of "negative entropy," i.e., the stuff that enables living systems to create and maintain order in the universe).[58] From the transhumanist point of view, the result would be, all things considered, no less tragic than if humanity were to go extinct.[59]

A key idea here is that the potential *value* of our posthuman future could be unimaginably huge. For example, one estimate suggests that a total of "a hundred thousand billion billion billion"—that is, a 1 followed by 32 zeros, or 100,000,000,000,000,000,000,000,000,000, 000—humans could someday populate the universe.[60] These people might colonize a large fraction of our future light cone, use enhancement technologies to radically augment their cognitive and moral capacities, live indefinitely long lives through rejuvenation therapies, upload their minds to achieve a kind of digital immortality, and perhaps even convert entire planets into supercomputers that run simulations in which conscious beings live happy, worthwhile lives (thereby increasing the total amount of well-being in the cosmos, which some ethical theories prescribe).[61] As Parfit puts the point, "Life can be wonderful as well as terrible, and we shall increasingly have the power to make life good. Since human history may be only just beginning, we can expect that future humans, or supra-humans, may achieve some great goods that *we cannot now even imagine.*"[62] In a phrase, the expected value of the future is *astronomically high* given the potential number and nature of our posthuman descendants. Let's call this the **astronomical value thesis**.[63]

This leads Bostrom to argue that "the loss in expected value resulting from an existential catastrophe is so enormous that the objective of reducing existential risks should be a dominant consideration whenever we act out of an impersonal concern for humankind as a whole." In other words, we should behave according to the following "rule of thumb for such impersonal moral action," dubbed **Maxipok**:

> *Maximize the probability of an "OK outcome," where an OK outcome is any outcome that avoids existential catastrophe.*[64]

One can think of our predicament as follows: the present moment—a century that the Long Now Foundation writes as "02000" to encourage "deep time" thinking—is a narrow foundation upon which an extremely tall skyscraper rests.[65] The entire future of humanity resides in this skyscraper, towering above us, stretching far beyond the clouds. If this foundation were to fail, the whole building would come crashing to the ground. Since this would be astronomically bad according to the above thesis, it behooves us to do everything possible to ensure that the foundation remains intact. The future depends crucially on the decisions we make today, just as the present depends on the decisions made by our ancestors, and this is a moral burden that everyone should feel pressing down upon their shoulders.[66]

While one might accept that every human perishing tomorrow would be an unthinkable catastrophe, one might also object that there is no particular reason to value the lives of people who do not yet exist. Why should current people care about generations that are born 100, 10,000, or even 100 million years from today? What obligations do we really have to future people in some far-off, exotic futureland? Many moral philosophers respond that *when* one exists should be irrelevant to that person's moral status. By analogy, *where* one exists should be— it appears correct to assert—irrelevant to whether or not one matters ethically: e.g., the suffering of a child in Johannesburg is just as bad as the suffering of a child in Copenhagen, Beijing, or Honolulu. And since modern physics reveals that space and time form a unified four-dimensional continuum (called "spacetime"), there don't appear to be any *fundamental* reasons for privileging one dimension over another, meaning that "affecting a temporally distant individual in the future is similar to affecting a spatially distant individual" right now.[67] If one rejects "space discounting" (or devaluing the lives of people who are spatially distant from us), one should also reject "time discounting" (or devaluing the lives of people who are temporally distant from us).

Furthermore, as the risk expert Jason Matheny observes, time discounting future lives yields conclusions that "few of us would accept as being ethical."[68] For example, if one were to discount future "lives at a 5% annual rate, a life today would have greater intrinsic value than a billion lives 400 years hence"—i.e., a single person dying this evening would constitute a *worse moral tragedy* than a global catastrophe that kills 1 billion people in four centuries.[69] Similarly, a 10 percent annual discount rate would entail that one person today is equal in value to an unfathomable 4.96×10^{20} people 500 years from now.[70] This line of reasoning appears to be not only misguided but *outrageously wrongheaded*, from which it follows that discounting human lives is deeply problematic.[71]

The futurist Wendell Bell offers seven additional reasons that contemporary generations have obligations to future generations. These are:

(1) *A concern for present people implies a concern for future people.* There is no "clear demarcation . . . between one generation and the next," meaning that "a concern for people living now carries us a considerable way into caring about future people." Imagine that you have children who have children. You care about your grandchildren, who will one day care about their own grandchildren. The result is an unbroken chain of caring that extends indefinitely into the future.

(2) *Thought experiments in which choosers do not know to which generation they belong rationally imply a concern for both present and future people.* If one knows nothing about which generation one will live and is asked "to choose how each generation ought to behave, consuming now or saving and preparing for the future," rational choosers will "allow for the well-being of both present *and* future generations." (This thought experiment borrows from John Rawls's idea of the "original position," in which people select principles upon which society will be based without knowing

anything about their gender, ethnicity, social status, and so on.[72])
It follows that "we ought to care about the well-being of future
people because that is what rational people would choose to do if
they did not know what generation they were in."

(3) *Regarding the natural resources of the earth, present generations
have no right to use to the point of depletion or to poison what they
did not create.* Since natural resources were not produced by any
human, "everyone has a right to their use, including members
of future generations." Therefore, "the members of the present
generation have an obligation to future generations of leaving the
earth's life-sustaining capacities in as good a shape as they found
them or of providing compensating benefits of life-sustaining
worth equal to the damage that they do."

(4) *Past generations left many of the public goods that they created not
only to the present generation but to future generations as well.* This
suggests that "no generation has the right to use up, totally consume, or destroy the existing human heritage, whether material,
social, or cultural, so that it is no longer available to future generations."

(5) *Humble ignorance ought to lead present generations to act with
prudence toward the well-being of future generations.* We are only
beginning to understand the universe, and we have only the vagu-
est sense of "what the human destiny is or might become." Thus,
"weighted with such ignorance, the present generation ought to
act prudently so as not to threaten the future survival and well-
being of the human species."

(6) *There is a prima facie obligation of present generations to ensure
that important business is not left unfinished.* The term "important
business" here refers to "human accomplishments, especially ex-
ceptional ones in science, art, music, literature, and technology,
and also human inventions and achievements of organizational
arrangements, political, economic, social, and cultural institutions, and moral philosophy." Both this and the previous point

clearly connect to the transhumanist goal of reaching new and better modes of being.

(7) *The present generation's caring and sacrificing for future generations benefits not only future generations but also itself.* One way to give life meaning is through engagement and altruistic sacrifice. In other words, "it is through being concerned for other people, both living and as yet unborn, that a person achieves self-enrichment and personal satisfaction." As Bell adds, "Genuinely caring about future generations and taking effective action to benefit their well-being are objective and rational answers to the contemplation of one's own death and the feelings of futility and despair it produces. Thus, we can strengthen ourselves by creating a community of hope."[73]

So, there are compelling reasons for caring about the well-being of future people and, therefore, allocating a nontrivial sum of resources for existential risk research. From a methodological standpoint, this is why the present book considers a wide range of risk scenarios, including some that have a *prima facie* "sci-fi" flavor: given the astronomically high stakes involved, even risky phenomena that seem, from a "pre-theoretic" perspective, unlikely warrant further investigation.[74] Perhaps future research will reveal certain scenarios to be less problematic than initially expected, in which case we can safely ignore them; but it might also show them to be *worse* than anyone imagined, thus requiring immediate action to curb a cataclysm. The only way to know is to put these ideas—all of them, despite any prior prejudices (see section 1.6)—under the electron microscope of critical analysis and to go from there. As Rees eloquently puts it in the foreword of this book, "The stakes are so high that those involved in this effort will have earned their keep even if they reduce the probability of a catastrophe by a tiny fraction."

1.5 Fermi and Filters

Let's now consider some general features of our place in the universe, beginning with the **Fermi paradox**. Named after the physicist Enrico Fermi, who worked on the Manhattan Project, this paradox originated during a 1950 luncheon conversation about the possibility of other civilizations populating the universe. After pondering the issue, Fermi exclaimed, "Where is everybody?" The reasoning goes like this: some 10 billion galaxies and 1 billion trillion stars exist in the observable universe. A certain percentage of these stars will likely have Earth analogs in the habitable or "Goldilocks" zone, the region around a star where conditions are suitable for liquid water and, therefore, carbon-based lifeforms. Given these facts, we should expect a large number of technologically advanced civilizations to exist—that is to say, *even if* the probability of an advanced civilization developing on any given exoplanet is minuscule, the sheer number of exoplanets in the cosmos should make advanced civilizations abundant.

Yet, dubious anecdotes and grainy footage aside, we see no legitimate signs of extraterrestrial life crying out for cosmic companionship in the darkness of space. We have encountered no aliens with imperialistic ambitions to dominate the galaxy. We find no rapacious swarms of von Neumann probes buzzing around us—that is, spacecraft capable of mining resources throughout the universe to create copies of themselves, thereby producing an exponential expansion of probes in all three dimensions. And we have detected no verifiable squeaks in the form of nonrandom electromagnetic signals washing up against our planetary island.[75] This is the Fermi paradox: the skies are silent when they should be noisy.

Or, perhaps there is a flaw in the above reasoning. In 1961, the astrophysicist Frank Drake proposed an "equation" that attempts to specify all the crucial variables that scientists must consider to calculate the total number of communicable civilizations in the universe. The result is the **Drake equation**, which states that $N = R^*$ x fp x ne x fl x fi x fc x L. These variables stand for the following:

N is the total number of communicable civilizations.

$R*$ is the rate of formation of stars suitable for the development of intelligent life.

fp is the fraction of those stars with planetary systems ("p" for planets).

ne is the number of planets, per solar system, with an environment suitable for life ("e" for ecologically suitable).

fl is the fraction of suitable planets on which life actually appears ("l" for life).

fi is the fraction of life bearing planets on which intelligent life emerges ("i" for intelligence).

fc is the fraction of civilizations that develop a technology that releases detectable signs of their existence into space ("c" for communicative).

And L is the length of time such civilizations release detectable signals into space.[76]

It is difficult to determine accurate values for each of these variables, and consequently estimates have varied dramatically. For example, Drake and others initially calculated that the number of civilizations in the Milky Way could range between 1,000 and 100 million. In *Cosmos*, Carl Sagan estimates that there are perhaps 1 billion planets in the universe that have harbored civilizations, but only about ten civilizations with the radio astronomy that would enable them to communicate with other civilizations. More recent calculations using low estimates for different variables suggest that we are alone in the universe ($N = 9.1 \times 10^{-11}$), while others using high estimates suggest that there could be more than 150 million advanced civilizations ($N = 1.5 \times 10^8$).

Either way, *observations* suggest that humanity is alone. The science fiction writer David Brin refers to this eerie situation of cosmic isolation—a kind of sensory deprivation—as the **Great Silence**.[77] Later, the economist Robin Hanson proposed an explanatory framework for the Great Silence, the central idea being that there must exist at least one **Great Filter** on the path from dead matter to advanced civilizations capable of communicating with other advanced civilizations. Hanson identifies nine major evolutionary transitions that have to obtain for a civilization to reach a communicable state:

(1) The right star system (including organics)

(2) Reproductive something (e.g. RNA)

(3) Simple (prokaryotic) single-cell life

(4) Complex (archaeatic and eukaryotic) single-cell life

(5) Sexual reproduction

(6) Multicellular life

(7) Tool-using animals with big brains

(8) Where we are now

(9) Colonization explosion.[78]

Perhaps the emergence of information-carrying, self-replicating molecules (such as ribozymes, also known as RNA enzymes) is the probability bottleneck that explains the Great Silence. After all, despite decades of research, scientists have failed to produce a single instance of abiogenesis ("life from non-life") in the laboratory, no matter how carefully they recreate the hypothesized geophysical conditions of our primordial planet. (Although some, such as Stanley Miller and Harold Urey, have managed to produce the constituents of proteins from inorganic compounds.) Or maybe the rise of intelligent tool-using animals with a high encephalization quotient (i.e., brain-to-body ratio) constitutes the Great Filter. As the biologist E.O. Wilson once suggested, "Perhaps one of the laws of evolution across inhabited planets in the universe . . . is that intelligence usually extinguishes itself."[79] There could also be *multiple* Great Filters between (1) and (9), with the limiting case being a Great Filter at each transition.[80]

The ultimate question for existential risk scholars is whether or not a Great Filter lies in our future. One way to evaluate this question is to look backward and consider how probable the steps before (9) are. If we find that (1) through (8) are reasonably likely, then we should conclude that a Great Filter probably lies in the future. In Hanson's words, "Optimism . . . regarding our future is directly pitted against optimism regarding the ease of previous evolutionary steps. To the extent those successes were easy, our future failure to [reach technological maturity] is almost certain."[81] This is precisely why Bostrom argues that discovering single-celled organisms on Mars, if independent in origin from those on Earth, would be a crushing disappointment: it would reduce the probability of one or more Great Filters associated with (1) to (3). Similarly, finding complex organisms capable of sexual reproduction would lower the probability of Great Filters associated with (1) to (5).[82] The result would be to "shift the probability more strongly to the hypothesis that the Great Filter is ahead of us."[83] By analogy, say that conditions A, B, and C are necessary and sufficient for X to obtain. If X is failing to obtain and you know that A is almost always the case, then A probably isn't the reason for X failing, so the probability that B or C is the obstructing factor increases. Thus, we should hope to find the universe utterly vacant, since this would suggest that the Great Filter lies somewhere in our past. As Bostrom wryly declares, "Dead rocks and lifeless sands would lift my spirit."[84]

On the other hand, imagine a future in which we build supercomputers capable of simulating our evolutionary history. Imagine that such simulations begin with a "lifeless" universe but that after running a large number of them we find primitive lifeforms evolving in a majority of the universes. Depending on how high-resolution the simulations are, we could take this to infer that step (1) is not improbable. Now imagine that these single-celled creatures consistently evolve into tool-using, big-brained organisms but almost never manage to establish industrial societies. What would this imply? If scientists were to find the simulated creatures consistently evolving to a particular step between (1) and (7) but not beyond, then we would have reason for thinking that the Great Filter lies behind us. In contrast, if many of our simulations were to yield industrial societies like ours but *not* technologically mature civilizations that emit powerful signals into the heavens, colonize some portion of their Hubble volume, or launch von Neumann probes into space, then we would have greater reason for worrying about a killer catastrophe up ahead.[85]

So, using this logic, the concept of the Great Filter can help clarify the degree to which contemporary people should be nervous about phenomena like climate change, biodiversity loss, nuclear weapons, biotechnology, synthetic biology, molecular nanotechnology, artificial intelligence, and so on. If science establishes that the evolutionary transitions behind us are relatively likely, then we should fear that doom lies in our future. (For more on the Great Filter framework and the probability of doom, see section 7.1.)

1.6 Biases and Distortions

Determining the extent to which we might be in danger requires precise and accurate thought about our evolving existential situation. Yet—at the risk of asserting a platitude—thinking clearly about the world is difficult. Our cognitive capacities are limited by the information-encoding and concept-generating mechanisms bequeathed to us by evolution and, as Bruce Tonn and Dorian Stiefel report, "most individuals' abilities to imagine the future goes 'dark' at the ten-year horizon."[86] Making matters worse, our minds are susceptible to a range of *cognitive biases* that can trick us into embracing—sometimes with great confidence—incorrect beliefs about reality. Given that the stakes are astronomically high, scholars should be especially careful to guard against the many intellectual prejudices that can distort our thinking. A short list of biases relevant to existential risk studies includes:

(i) **Conjunction fallacy.** Consider Linda, who "is 31 years old, single, outspoken, and very bright. She majored in philosophy. As a student, she was deeply concerned with issues of discrimination and social justice, and also participated in anti-nuclear demonstrations."[87] Given this information, which of the following two statements is more *probable*: (a) Linda is a bank teller, or (b) Linda is a bank teller and is active in the feminist movement? When subjects are asked this question, the majority opt for (b) over (a). After all, (b) is more *representative* of Linda's description, and consequently it appears more plausible. But plausibility does not equal probability. In fact, the objectively correct answer is that (a) is more likely true than (b). Why? Because (b) contains (a), resulting in an asymmetry such that for (b) to be true, (a) must also be true, but for (a) to be true, (b) need not be true.[88]

Anytime a proposition is added to another proposition, the resulting conjunction is (as a whole) necessarily less probable than either of the two propositions individually.[89] This is because two propositions conjoined and asserted as true require more to be the case in the world (assuming the *correspondence theory of truth*).[90] Whereas (a) requires one condition to hold, (b) requires two. This loosely relates to the principle of Occam's razor, which states that when two hypothesis explain a given phenomenon, or *explanandum*, equally well, one should always choose the simpler hypothesis. As the philosopher Graham Oddie writes in the *Stanford Encyclopedia of Philosophy*, the "degree of informative content varies inversely with probability—the greater the content the less likely a theory is to be true."[91]

Now consider an alternative situation involving Linda. Which of the following is more probable: (a) Linda is active in the feminist movement, or (b) Linda is active in the feminist movement *or* is a bank teller? The correct answer now is that (b) is more probable, since it could be true *even if* Linda isn't active in the feminist movement. In other words, (b) increases the number of ways that it could be true by adding another proposition not through conjunction but disjunction. And the more that disjuncts are added, the more probable the resulting proposition (as a whole) will be.[92]

The relevance is this: existential risk scenarios like human extinction and permanent stagnation are causally *disjunctive*. That is, they could happen as a result of asteroids *or* supervolcanoes *or* climate change *or* nuclear war *or* designer pathogens *or* superintelligence, etc. Yet the human mind "prefers" conjunctions. Consequently, we may overestimate elaborate risk scenarios while underestimating the total risk posed by a growing number of deadly threats, or we may judge elaborate arguments against certain risk scenarios to be more convincing than they are, which could leave us unnecessarily vulnerable.

(ii) **Confirmation bias.** John is a huge supporter of a politician named Zoe. Unfortunately, his close friends don't share his excitement because they believe that Zoe is a pathological liar. To convince himself that Zoe is trustworthy, John curates ten impressive instances when Zoe told a hard truth, complete with verifiable citations (e.g., from PolitiFact). Does this evidence justify his prior beliefs about Zoe's probity? No, because evaluating truth-claims requires taking into account both confirming *and* disconfirming cases—an issue we will revisit in the next section. There could, indeed, be 100 cases of Zoe offering a complete fabrication to cover up criminal acts and malfeasance, which would suggest that Zoe is duplicitous after all. The flip side of this phenomenon is the disconfirmation bias, which occurs when one spends more time scrutinizing evidence that contradicts one's preferred beliefs than evidence that supports them. For example, imagine that John's friends present the 100 instances of Zoe lying to John in an attempt to sway his opinion. Since John wants to believe that Zoe is truthful, he responds by assiduously researching every single accusation to show how each might be flawed. In contrast, he spends virtually no time ensuring that the ten instances of Zoe stating the facts are accurate beyond a reasonable doubt.

This bias could nontrivially influence work on existential risks. For example, a stubborn optimist might spend all her time poking holes in arguments that humanity is in danger while uncritically elevating data that suggests our future is safe. The result of such tendentious research, on the optimist's part, could have catastrophic consequences if she were to persuade society to let down its guard. Alternatively, an existential risk scholar with alarmist inclinations and a career predicated on there being a high threat level might employ the exact same techniques to reach exaggerated conclusions about how risky our situation is. Both cases must be avoided, and the only way to do this is to embrace the epistemic attitude of *intellectual honesty*, which means (a) considering all the evidence, and (b) treating all the evidence the same, even when this leads to psychological disappointment.

(iii) **Observation selection effect**. This is a type of selection effect that arises from the fact that certain types of catastrophes are incompatible with the existence of observers like us. It can lead people to overestimate the probability of survival based on the empirical fact that human extinction has never before occurred. But observers like us can only ever find themselves in situations in which there are no extinction events in our species' evolutionary past. Thus, the fact that we have not yet gone extinct should not be surprising. Similarly, consider an Ultimate X-Risk that could destroy the entire universe in an instant. Can the past provide any useful information about how probable this event is? Apparently not. Whether or not an Ultimate X-Risk is extremely probable or improbable, we should expect to find ourselves in a world exactly like this one, with fully intact galaxies, stars, and planets. Both hypotheses (probable versus improbable) predict the very same observations. As Ćirković puts the point, "People often erroneously claim that we should not worry too much about existential disasters, since none has happened in the last thousand or even million years. This fallacy needs to be dispelled."[93]

Other cognitive distortions relevant to existential risk studies include:

- **Availability bias**: This occurs when people "rely too strongly on information that is readily *available* [while ignoring] information that is less available."[94]

- **Gambler's fallacy**: "The tendency to think that future probabilities are changed by past events, when in reality they are unchanged."[95]

- **Good-story bias**: Our intuitions about the future are often shaped by popular books and movies, and thus may be biased toward exciting storylines, independent of their probability.[96]

- **Affect heuristic**: This "refers to the way in which subjective impressions of 'goodness' or 'badness' can act as a heuristic, capable of producing fast perceptual judgments, and also systematic biases."[97]

- **Motivated reasoning**: "Rather than search rationally for information that either confirms or disconfirms a particular belief, people actually seek out information that confirms what they already believe."[98]

- **Scope neglect**: This "occurs when the valuation of a problem is not valued with a multiplicative relationship to its size."[99]

- **Superiority bias**: "The belief that you are better than average in any particular metric."[100]

- **Negativity bias**: The human tendency to react more strongly to stimuli that have a negative valence.

- **Optimism bias**: The persistent belief that the future will be better than the past and present.[101]

- **Anchoring**: "The common human tendency to rely too heavily on the first piece of information offered (the 'anchor') when making decisions."[102]

- **Base rate fallacy**: This happens when "people order information by its *perceived degree of relevance*, and let high-relevance information dominate low-relevance information."[103]

- **Hindsight bias**: "A memory distortion phenomenon by which, with the benefit of feedback about the outcome of an event, people's recalled judgments of the likelihood of that event are typically closer to the actual outcome than their original judgments were."[104]

- **Overconfidence**: This involves someone believing "that his or her judgement is better or more reliable than it objectively is."[105]

Although we won't discuss these any further here, readers are strongly encouraged to familiarize themselves more closely with these phenomena.[106]

1.7 The Epistemology of Eschatology

Eschatology: the study of the end of the world
Epistemology: the theory of knowledge

Just as section 1.5 placed humanity in a larger *cosmic* context, let's now consider the broader *cultural* context in which we find ourselves. This section makes several important points that underlie and motivate the nascent field of existential risk studies, and although it may appear to delve into excessive detail, I would encourage readers not to dismiss this material too quickly.

To begin, the record of human beings claiming that their generation is the last is historically extensive—far more extensive than is generally known. The first linear **eschatological** narrative was probably invented by the ancient Persians. According to the prophet Zoroaster, also known as Zarathustra, cosmic history consists of three or four periods (depending on the tradition), each of which is exactly three millennia long. The last period culminates with the arrival of a messianic virgin-born savior, the Saoshyant, who will usher in a bodily resurrection of the dead, a Final Judgment of humanity, and an Armageddon-like war between the cosmic opposites of Good and Evil. This eschatology very likely influenced the end-times narratives of Judaism (during the Second Temple period), and consequently the two other Abrahamic religions, namely, Christianity and Islam. If this is true, which appears to be the case, then we have an argument for Zoroaster being *the most influential human to have ever existed.*[107]

Now, consider how the popular interpretation of Christian scripture known as *dispensationalism* compares to the above, albeit brief, story. According to this view, history consists of seven distinct periods called "dispensations." Contemporary humans are living in the second-to-last dispensation known as "Grace," which will conclude after Jesus briefly returns to Earth to "rapture" all the Christians, both alive and dead, who have existed since roughly 70 CE.[108] After this, a seven-year period called the Tribulation will commence, during which the Antichrist will rule a powerful governmental body like the European Union or the United Nations.[109] People will suffer immensely, especially the Jews and those who convert to Christianity after the rapture. The end of the Tribulation will be marked by the Second Coming of Christ (the *Parousia*) and the battle of Armageddon, perhaps in propinquity to the ancient town of Megiddo, Israel. Jesus will cast the Antichrist into the Lake of Fire, and the final dispensation—the Millennial Kingdom—will commence.[110] At the end of this 1,000-year period, there will be yet another great battle, this time between God and Satan, involving the nations of Gog and Magog, followed by another bodily resurrection of the dead and one last judgment of humanity, called the "Great White Throne Judgment."[111] All true Christians will enter paradise in heaven and the unbelievers will be banished to perdition for eternity.

Paralleling this narrative in certain notable respects, some traditions in Sunni Islam prophesy that an end-of-days messianic figure called the Mahdi will appear in Mecca, Saudi Arabia, and lead an army of Muslims into an Armageddon-like battle in the small town of Dabiq in northern Syria, near Aleppo. After Armageddon, the remaining Muslim army will travel to and supernaturally conquer Constantinople (now Istanbul).[112] The Antichrist, or *Dajjal*, will then make his appearance, spreading horrible evil throughout the entire world. But his arrival, the first of Ten Major Signs of the Last Hour, will be followed by the second Major Sign, namely, the descent of Jesus on the wings of two angels. This will occur over the White Minaret of the Umayyad Mosque, in modern-day Damascus. Jesus will then chase the Antichrist to the "gate of Ludd," now called "Lod" in Israel, at which point he will kill the Antichrist. Other Major Signs will follow, most of which are quite bizarre, such as the sun rising from the West and the emergence of the ferocious killing machines Gog and Magog, whom God will utterly decimate. At the very terminus of cosmic history, God will oversee a bodily resurrection and Final Judgment of humanity. All true Muslims will enter heaven and the infidels will be cast into hell forever.[113]

There are a couple of issues worth pausing over here. First, I would argue that it is vital for existential risk scholars to understand these narratives in some detail. The reason is that they are widely believed around the planet and have shaped world history in *truly pro-*

found ways. Consider that an incredible 41 percent of U.S. Christians in 2010 avowed that Jesus will either "definitely" or "probably" return by 2050.[114] One finds a similar prevalence of end-times beliefs in the Muslim world, with, for example, 83 percent of Muslims in Afghanistan and 72 percent in Iraq claiming that the Mahdi will return within their lifetimes.[115]

Looking back to the origin of these faiths, both Jesus and Muhammad may have believed that the end was nigh in their own day. As the majority of New Testament scholars today maintain—following the influential theologian Albert Schweitzer—Jesus was probably a failed apocalyptic prophet who voluntarily sacrificed himself "to force the hand of God" when it became clear that the world was not about to end.[116] With respect to Islam, the historian Allen Fromherz writes that "some scholars have suggested that Islam was, from the first revelations of Muhammad, almost entirely an apocalyptic movement.... Some have even supposed that Muhammad deliberately failed to designate a successor because he predicted that the final judgment would occur after his death."[117]

Furthermore, numerous conflicts of historical significance have been greatly influenced by interpretations of Christian and Islamic eschatology—a phenomenon that I call the "clash of eschatologies."[118] For instance, as subsection 4.3.1 explores, many contemporary Islamic terrorist groups, both Sunni and Shia, are animated by "active apocalyptic" beliefs according to which they see themselves as *fervent participants in an apocalyptic narrative that is unfolding in real-time.*[119] But the plot thickens, because some of the most prominent Islamic terrorist groups today have emerged in direct response to two recent U.S.-led incursions, namely, the 1990 Gulf War and the 2003 Iraq War. And both of these may have been shaped by eschatological convictions associated with what scholars call the "Armageddon lobby" in the United States—that is, a large demographic of leaders and constituents whose political worldviews are intimately linked to dispensationalism.[120] Even more, many Islamists accuse Western forces stationed in the Middle East of being "crusaders," a term that gestures back to the religious wars of the Crusades; and as the terrorism expert Will McCants notes, "The 100,000 European foreign fighters who flooded into Palestine under the banner of the First Crusade believed they were hastening the End of Days."[121] So, the ongoing violence in the Middle East—currently the world's epicenter of conflict—has been fueled *for centuries* by end-times beliefs held by both Christians and Muslims.

Perhaps most intriguingly, the two most consequential "secular" movements of the twentieth century, namely, Marxism and Nazism, appear to have been inspired by religious grand narratives of history. For example, Marx believed that humanity started out in a state of primitive communism (the Garden of Eden), after which we passed through stages (dispensations) like feudalism and capitalism. In the end, humanity will enter into a paradisiacal world of pure communism (heaven on Earth) thanks to the efforts of Marx (a messianic prophet), who introduced the message of communism to the proletariat. But this last step to paradise will only occur, as the historians Daniel Chirot and Clark McCauley note, after "a final, terrible revolution" (Armageddon) that will "wipe out capitalism, alienation, exploitation, and inequality" (sin).[122] Similarly, Chirot and McCauley write that

> *It was not an accident that Hitler promised a Thousand Year Reich, a millennium of perfection, similar to the thousand-year reign of goodness promised in Revelation before the return of evil, the great battle between good and evil, and the final triumph of God over Satan. The entire imagery of his Nazi Party and regime was deeply mystical, suffused with religious, often Christian, liturgical symbolism, and it appealed to a higher law, to a mission decreed by fate and entrusted to the prophet Hitler.*[123]

It is considerations like these that lead the biblical scholar and terrorism expert Frances Flannery to declare that "the Book of Revelation has arguably been responsible for more genocide and killing in history than any other [book]." Elsewhere she claims that Revelation is

*responsible, directly or indirectly, for massive amounts of vio-
lence. In fact, it is arguably the bloodiest book in history. Even
today, groups and individuals as diverse as the Oklahoma City
bombers and radical Islamist groups . . . have each updated the
Book of Revelation to apply to their own period and causes, using it to justify violence and brutality.*[124]

Thus anxious anticipation of, and even outright elation about,
the apocalypse can be found across cultural space and time.[125] This
leads to a second important point: the fact that so many people have
sounded the alarm bell throughout history may lead some observers to dismiss contemporary concerns from the existential risk community about global catastrophic risks. Such skeptical people might
say, "Why should I believe doomsaying scientists? *Every* generation
throughout history has had *somebody* claiming that their generation
is the last. This is just more of the same alarmist nonsense."

But this objection is deeply misguided for reasons relating to a
single crucial topic: **epistemology**. This refers to the subfield of philosophy dedicated to understanding truth, justification, and knowledge. Epistemological questions include: What constitutes truth?
What conditions make a belief reasonable? Of what does knowledge
consist? The most important issue for the present discussion concerns
what we can call "epistemic *justification, warrant,* or *reasonableness,*"
where these terms are more or less interchangeable in this context.

The point is that *science*—our very best strategy for acquiring
knowledge about the universe—is based on a highly rigorous interpretation of epistemic justification. Theories must be not merely
compatible with, but positively *supported by* some form of intersubjectively verifiable evidence.[126] And not just any evidence, but rather
the *totality of evidence available at a given time.*[127] This last point is
important for the following reason: imagine two competing hypotheses, A and B. Hypothesis A has, let us say, two "pieces" of evidence
supporting it. Should one accept it as true, given this evidential support? The answer depends on whether hypothesis B has more than,
less than, or equal to two "pieces" of evidence. If B has, for instance, 20
"pieces" of evidence in its favor, then it would be irrational to believe

A. The *totality condition* of reasonable belief is a feature that many religious extremists, conspiracy theorists, and psychotic people fail to consider, thus leading them to accept unwarranted propositions that nonetheless may have some evidential support. Since humanity can't peek under the hood of reality, the best we can hope for in life is to be as reasonable as possible—that is, to construct worldviews whose interlocking beliefs are founded on objective evidence considered as a whole and constantly responsive to changes in the pool of available evidence as ongoing research uncovers new data.[128]

The further point is that, as indicated above, existential risk studies is a thoroughly *scientific* discipline. It uses the tools and methods of *rational empiricism* to map out the obstacle course of risks that civilization must navigate in the coming decades and centuries—and beyond. Even in the case of highly speculative risk scenarios such as a superintelligence takeover or a simulation shutdown, the core line of reasoning involves empirical trends, objective knowledge, and logical inferences. In contrast, the world's many religious traditions are based not on evidence but faith, and the source of knowledge comes not from observation but private revelation and testimony.[129] This makes the epistemological status of religious eschatology *fundamentally incommensurable* with that of existential risk studies, and this difference accounts for why one should listen to scientists and philosophers worried about the apocalypse but not religious folks.

To adapt a phrase from the philosopher David Hume, *the wise person always proportions her or his fears to the best available evidence, considered as a whole.* It follows that fear itself is not bad or undesirable as long as it is rational. Indeed, our best chance of surviving this century is to let what we might call *intelligent anxiety* be our guide and chaperone as we move forward. Just as long as this anxiety is motivating rather than defeatist (see section 7.2), it could be the key that unlocks our posthuman future.

Chapter 2: Our Cosmic Risk Background

2.1 Threats from Above and Below

The astrophysicist Neil deGrasse Tyson was once asked before filming a video for Big Think to briefly discuss any topic of his choosing. In deadpan fashion, Tyson intoned that "the universe is a deadly place. At every opportunity, it's trying to kill us. And so is Earth."[1] Humor aside, this gestures at a truth about our existential situation: the universe is an obstacle course of deadly hazards, and it doesn't care whether intelligent life survives or perishes. We can call this obstacle course our **cosmic risk background**. There are two general risk types within this category, namely, (a) those emerging from Earth, and (b) those hiding in the heavens. Supervolcanoes and natural pandemics are examples of the former, whereas asteroids, comets, and other astronomical phenomena are instances of the latter. Let's examine these in turn.

2.2 Supervolcanoes

To review some common geological knowledge, a volcano is an opening in Earth's surface through which magma and the dissolved gases that it contains escape—sometimes violently. Scientists have devised the volcanic explosivity index (VEI) to classify the strength of eruptions. The VEI ranges from 0 to 8, where the continuous volcanic flows on Hawaii with relatively small eruptive volumes and plume heights of less than roughly 330 feet constitute a 0 and the 1815 erup-

Table A. Volcanic Explosivity Index with Examples

VEI	Examples
8	**Toba, 72,000 BCE; Yellowstone, 640,000 BCE** "Mega-colossal" with "vast" stratospheric injections
7	**Tambora, 1815** "Super-colossal" with "substantial" stratospheric injections
6	**Pintabu, 1991** "Colossal" with "substantial" stratospheric injections
5	**Vesuvius, 79** "Paroxysmic" with "significant" stratospheric injections
4	**Calbuco, 2015** "Cataclysmic" with "definite" stratospheric injections
3	**Nabro, 2011** "Catastrophic" with "possible" stratospheric injections
2	**Sinabung, 2010** "Explosive" with no stratospheric injections
1	**Stromboli, continuous** "Gentle" with no stratospheric injections
0	**Kīlauea, ongoing** "Effusive" with no stratospheric injections

tion of Mount Tambora, located on the Indonesian island of Sumbawa, constitutes a 7. (See Table A.) Let us linger on the latter for a moment. On April 5, 1815, Mount Tambora began to spew ash into the air. Subsequent explosions were loud enough for soldiers hundreds of miles away to wonder if a war might have broken out. Five days later, a plume of smoke reached 25 miles high, propelled by three pillars of fire that eventually merged into a single column of blazing rock. Toxic ash and pumice almost eight inches wide rained down upon Sumbawa, and a tsunami crashed into the beaches of nearby islands. Dead vegetation entangled with buoyant pumice created massive "rafts" floating on the ocean, some over three miles across. An estimated 10,000 people on the island died instantly from the blast, while many more perished in the aftermath, due to starvation and disease. In fact, the word "Tambora" means "gone" in the local language.[2]

But the worst effects were those observed across the Northern Hemisphere a year later, during the summer of 1816. Throughout Europe, the U.S., and Asia, unusually cold weather ruined the year's crops, leading to widespread food shortages. In France, this resulted in rioting; in Ireland, where rain fell for eight weeks without a hiatus, famine and malnutrition brought about an outbreak of typhus that killed thousands; in Bengal, an epidemic of cholera emerged that, after spreading around the globe, caused tens of millions of deaths; in China, people starved and some parents even killed their children "out of mercy"; and in the United States, ice covered lakes and snow blanketed regions of the East Coast as far south as Virginia during June and July.[3] This appears to have spurred a migration of folks from the U.S. Northeast into the American heartland, as Robert Evans notes in a *Smithsonian* article:

> *Odd as it may seem, the settling of the American heartland was apparently shaped by the eruption of a volcano 10,000 miles away. Thousands left New England for what they hoped would be a more hospitable climate west of the Ohio River. Partly as a result of such migration, Indiana became a state in 1816 and Illinois in 1818.*[4]

Perhaps most intriguingly, the anomalous weather inspired a then-unknown author named Mary Shelley, vacationing in Switzerland with the British poet Lord Byron, to write *Frankenstein*.[5] Lord Byron himself composed a poem in July of 1816 called "Darkness," which includes the lines "I had a dream, which was not all a dream. / The bright sun was extinguish'd, and the stars / Did wander darkling in the eternal space, / Rayless, and pathless, and the icy earth / Swung blind and blackening in the moonless air." This "Year Without a Summer" clearly illustrates how a large volcanic eruption can have major disruptive effects around the world.

But recall that there is one level higher on the VEI scale. This is reserved for **supervolcanic eruptions** capable of ejecting *hundreds of* *times* more ash into the atmosphere than Tambora did. When such an eruption occurs, sulfur dioxide is catapulted into the *stratosphere*, an atmospheric layer located above the troposphere and below the mesosphere, where the sun's light converts it into sulfuric acid. It then condenses into a layer of sulfate aerosols that reflect incoming solar radiation back into space, thereby causing Earth's skies to dim and surface temperatures to drop. The reduced photosynthesis from less sunlight can precipitate major agricultural failures lasting for years or even decades, resulting in, as the geologist Michael Rampino puts it, "widespread starvation, famine, disease, social unrest, financial collapse" and, at the extreme, "severe damage to the underpinnings of civilization."[6] Scientists refer to this scenario as a **volcanic winter**.

Numerous supervolcanic eruptions have occurred across geological time, at least 47 of which were known to science as of 2004.[7] One of the most recent happened on the Indonesian island of Sumatra circa 73,500 BCE—in fact, volcanologists coined the term "supereruption" to describe this particular event, known as the "Toba catastrophe." It may have led to a decade of severe weather changes, with average surface temperatures in the Northern Hemisphere falling by an incredible 5.4 to 9 degrees Fahrenheit.[8] According to Rampino, up to "three-quarters of the plant species in the Northern Hemisphere perished," and other studies suggest a spike in species extinctions at the time.[9] Even more, the Toba catastrophe may have caused a severe bottleneck in the population of our ancestors, with some experts estimating as few as 500 breeding females surviving, and human population sizes shrinking to "as small as 4000 for approximately 20,000 years."[10] Thus, if the diachronic tape of anthropological history were rewound and played again, *Homo sapiens* might not have made it through the Pleistocene.

On average, supereruptions occur about once every 50,000 years. As the Geological Society of London writes, "Sooner or later a supereruption will happen on Earth and this is an issue that . . . demands serious attention."[11] Unfortunately, our ability to predict supervolcanic eruptions is quite poor. For example, despite "2,000 years of observations for the Italian volcano Vesuvius, and a long history of monitoring and scientific study, prediction of the timing and magnitude of the next Vesuvian eruption remains a problem."[12] Similarly, Yellowstone National Park has seen three supereruptions over the past 2 million years, each of which "produced thick ash deposits over the western and central United States."[13] Recent studies show that the magma chamber under Yellowstone is 2.5 times bigger than previously thought, making it "close to the size of the pocket when the supervolcano last erupted, 640,000 years ago." The geoscientist James Farrell thus notes that "what we're seeing now agrees with the geologic data that we have about past eruptions. And that means there's the potential for the same type of eruption that we've seen in the past."[14] Yet we have no way of saying when this might happen.

But even if scientists *could* make accurate predictions, this might not help us *prevent* a supereruption from occurring. As the Geological Society of London observed in 2004, "Even science fiction cannot produce a credible mechanism for averting a super-eruption. The point is worth repeating. No strategies can be envisaged for reducing the power of major volcanic eruptions."[15] However, this may not be entirely true today. According to the GCR expert Seth Baum, one possible strategy involves drilling "the ground around potential supervolcanoes to extract the heat, although the technological feasibility of this proposal has not yet been established."[16] He adds that "this could be a very costly project, but, if it works, it could . . . reduce supervolcanoes GCR."[17] Either way, the point remains that prophylactic measures are highly limited. Perhaps our best chance of survival stems from post-eruption adaptation rather than pre-eruption mitigation, an issue to which we will return in section 6.5.

Although supervolcanoes rarely become active, spewing their innards high up into the atmosphere, they warrant serious concern because of the spatiotemporal scope of their consequences. If a Toba-sized supereruption were to occur tomorrow, the result could be a global or even existential catastrophe.

2.3 Natural Pandemics

Some scholars claim that the history of civilization is the history of war. While the amount of self-inflicted human suffering is truly stag-

Table B. Number of Deaths in Various Wars

Event	# of Deaths
World War II	85,000,000
Taiping Rebellion	35,000,000
Mongol conquests	30,000,000
World War I	21,000,000
Napoleonic Wars	7,000,000
Vietnam War	3,000,000
American Civil War	1,000,000
2003 Iraq War	500,000
War of 1812	24,000
Total	182,524,000

Note: Based on higher estimates of all these conflicts

gering, the facts suggest that infectious diseases have thrown more people into the grave than the innumerable conflicts fought over religion, ideology, resources, and pride. Consider the fact that from 1918 to 1920 the Spanish flu outbreak killed some 50 million people, whereas "only" about 17 million people died in World War I, which lasted from 1914 to 1918. Or note that about 3 percent of the global population (in 1940) died in World War II, whereas the Plague of Justinian killed roughly 50 percent of the European population at the time (beginning in the mid-sixth century).[18] Even more striking, the Black Death of Europe and Asia may have killed a total of 200 million people, which is more than the number of deaths caused by World

War II, World War I, the Mongol conquests, the Napoleonic Wars, the Vietnam War, the American Civil War, the 2003 Iraq War, and the War of 1812 *combined* (see Table B).

Consequently, infectious diseases like the flu, bubonic plague, and malaria have shaped world history in many important ways. For example, disease was a major factor behind the decimation of Native American populations after the arrival of Europeans, who, like most "civilized" peoples compared to their "primitive" counterparts, carried a much higher disease burden.[19] Similarly, smallpox played a role in enabling the Spanish to conquer the Aztec Empire, with it killing some 200,000 people in total and up to 75 percent of the population in some regions.[20] The Black Death in Europe remained a public health hazard for three centuries, "with a lasting impact on the development of the economy and cultural evolution."[21] And the HIV/AIDS pandemic from 1981 to 2006 may have snuffed out up to 65 million lives around the world—not to mention the socially harmful backlash against homosexuals from religious conservatives. More than any other infectious disease, though, malaria—caused by a parasitic protozoan and spread by the flying hypodermic needles called mosquitoes—has arguably had the greatest effect on humanity. According to the World Health Organization (WHO), about half the world's population today remains vulnerable to malaria, and during 2015 alone, some 214 million people contracted this disease, resulting in ~438,000 fatalities.[22]

It is important to note that most of the deaths caused by infection throughout history have been the result of *extreme outbreaks*. As the Global Challenges Foundation writes, "Plotting historic epidemic fatalities on a log scale reveals that these tend to follow a power law with a small exponent: many plagues have been found to follow a power law with exponent 0.26." The report adds that "if this law holds for future pandemics as well, then *the majority of people who will die from epidemics will likely die from the single largest pandemic.*"[23]

So, what reason do we have for expecting a pandemic to occur in the foreseeable future? Improvements in sanitation have significantly reduced the average person's exposure to pathogens, and modern medicine—in particular, vaccines and antibiotics—offer effective ways to prevent and treat infectious bugs. There are also international organizations like the WHO keeping a close and constant eye on disease outbreaks to minimize their impact, as demonstrated by the relatively successful containment of SARS and Ebola during the 2003 and 2014 epidemics, respectively. Yet these facts are counterbalanced by modern transportation systems that enable germs to travel from one continent to another at literally the speed of a jetliner, as well as dense urban areas like slums and megacities that make it far easier for pathogens to propagate through a population. In fact, the United Nations predicts "that 66% of the global population will live in urban centers by 2050."[24] Climate change will also exacerbate the risk of pandemics, since heat waves and flooding events will bring "more opportunity for waterborne diseases such as cholera and for disease vectors such as mosquitoes in new regions." Considerations like these have led many public health experts to claim that "we are at greater risk than ever of experiencing large-scale outbreaks and global pandemics," and that "the next outbreak contender will most likely be a surprise."[25]

There are also doctor-caused, or *iatrogenic*, illnesses that could become worrisome in the future, primarily for GCR reasons.[26] As the biomedical scientist Edwin Kilbourne writes, "An unfortunate result of medical progress can be the unwitting induction of disease and disability as new treatments are tried for the first time. Therefore, it will not be surprising if the accelerated and imaginative devising of new technologies in the future proves threatening at times."[27] Consider that in the United States alone "the true number of premature deaths associated with preventable harm to patients [is] estimated at more than 400,000 per year."[28] To put this in perspective, about 595,000 Americans were projected to have died of cancer in 2016— meaning that mistakes by doctors constitute a *major cause* of death.[29] If, as Kilbourne suggests, the medical sciences advance at an accelerating (perhaps exponential) rate, iatrogenic illnesses could become even more of a problem.

Another medicine-related threat stems from **superbugs**. This refers to *multidrug-resistant bacteria*, or bacteria that can't be treated using two or more antibiotics.[30] This has global risk implications because "antibiotics are the foundation on which all modern medicine rests. Cancer chemotherapy, organ transplants, surgeries, and childbirth all rely on antibiotics to prevent infections. If you can't treat those, then we lose the medical advances we have made in the last 50 years."[31] According to the Centers for Disease Control and Prevention (CDC), approximately 2 million people become sick as a result of superbugs each year, and some 23,000 die; but these numbers could be dwarfed by a global superbug outbreak. As the director general of the WHO Margaret Chan ominously puts it, "Antimicrobial resistance poses a fundamental threat to human health, development and security."[32]

Predicting a pandemic is extremely difficult; nonetheless, future global outbreaks are, it appears, more or less inevitable. As one commentator writes, "Experts say we are 'due' for one. When it happens, they tell us, it will probably have a greater impact on humanity than anything else currently happening in the world."[33]

2.4 Asteroids and Comets

At least one of the biggest extinction events on Earth was the result of an asteroid or comet collision. This occurred about 65 million years ago when an object ~10 kilometers across crash-landed on the Yucatan Peninsula, resulting in the extermination of all non-avian dinosaurs—an event that changed the trajectory of life by opening up new ecological niches for mammals.[34] An asteroid or comet might also have caused the devastating Permian-Triassic extinction some 251 million years ago (although some research indicates supervolcanism as the "kill mechanism"). This was the worst extinction event in planetary history, with "95 percent of all species, 53 percent of marine families, 84 percent of marine genera, and an estimated 70 percent of land species such as plants, insects and vertebrate animals" having perished.[35] There is, indeed, a startling record of large heavenly bodies wreaking mass havoc on Earth's biosphere. (See Box 3.)

As of this writing, scientists know about exactly 1,771 **potentially hazardous asteroids** circling Earth.[36] Such objects could, by definition, obliterate a sizable region of the planet, wiping out entire cities or coastlines. For example, if an asteroid were to descend above a

> **Box 3.** Consider a few recent close calls, beginning with the 2013 "Chelyabinsk event." This unfolded when an asteroid moving at about 42,000 miles per hour entered the atmosphere above the Russian city of Chelyabinsk, producing more light than the sun as it burned up. Numerous dashcam videos recorded the event, which damaged buildings, shattered windows, and injured nearly 1,500 people, resulting in 33 million U.S. dollars' worth of destruction. Four decades earlier, in 1972, a meteoroid "bounced" off Earth's atmosphere over the western United States, similar to the way a stone can skip across water. It came within 35 miles of Alberta, Canada, and if it had struck North America in the middle of the Cold War (note: a situation that the United States may be re-entering with Russia today) it could have initiated a retaliatory nuclear strike from the United States. This is known as the "Great Daylight Fireball." Looking back even further, an asteroid between 200 and 620 feet wide exploded over Siberia in 1908 with the energy output of a hydrogen bomb, flattening an area of forest roughly 770 square miles. Fortunately, the "Tunguska event" occurred over a region that was sparsely populated, so no one was injured.

high-density urban center, the resulting losses could be similar to the detonation of a nuclear weapon. In the latter case, even a relatively small impact could "on the more pessimistic analyses lead to waves 4–7 [meters] high all around the [Pacific] rim, presumably with the loss of millions of lives," since "over 100 million people live within 20 m of sea level and 2 km from the ocean."[37]

For an impactor to destroy civilization or bring about our extinction, though, it would need to be at least 1 kilometer across. Objects this large only strike Earth on average once every 500,000 years. If such a collision were to occur, it would kick up huge quantities of hot ash and dust into the stratosphere that would spread around the globe, blocking out incoming solar radiation. Consequently, "continental temperatures would plummet, and heat would flow from the warmer oceans onto the cooled land masses, resulting in violent, freezing winds blowing from sea to land."[38] An even higher-energy collision could bring about a global mass extinction event that would leave an indelible mark of catastrophe in fossiliferous strata. The astronomer William Napier describes this nightmare scenario as follows:

> *Regionally, the local atmosphere might simply be blown into space. A rain of perhaps 10 million boulders, metre sized and upwards, would be expected over at least continental dimensions . . . . Major global effects include wildfires through the incinerating effect of dust thrown around the Earth; poisoning of the atmosphere and ocean by dioxins, acid rain, sulphates and heavy metals; global warming due to water and carbon dioxide injections; followed some years later by global cooling through drastically reduced insolation, all of this happening in pitch black. The dust settling process might last a year to a decade with catastrophic effects on the land and sea food chains.*[39]

Indeed, the most commonly discussed risk associated with a large asteroid or comet impact is the possibility of an **impact winter**, similar to the volcanic winter phenomenon discussed above. This would induce global agricultural failures, mass starvation, malnutrition, and infectious disease outbreaks, all of which could cause major disruptions in the social, political, and economic foundations of civilization. At the extreme, an impact winter lasting years or decades could bring about a planetary-scale cataclysm from which humanity might never fully recover. As Napier concludes, in sobering language, "A great earthquake or tsunami may take 100,000 lives; a great impact could take 1,000 or 10,000 times as many, and bring civilization to an abrupt halt."[40]

2.5 Other Threats

The tapestry of cosmic risks that threaten our species—including supernovae, gamma-ray bursts, galactic center outbursts, superstrong solar flares, and black hole mergers or explosions—is vast, but most of these scenarios are exceptionally improbable. Thus, we will glance over only a few such risks here.[41]

(a) *Supernovae.* This occurs when a massive star uses up its nuclear fuel and thus cannot maintain its core temperature. The loss of thermal pressure pushing outward results in its core violently collapsing inward due to gravity, thereby producing a neutron star or stellar black hole. In the process, the imploding star releases a tremendous amount of energy. Another possibility is that a white dwarf in a binary star system (that is, locked in an orbital dance with another star) gains matter from its companion, eventually leading it to become so massive that its core collapses under its own gravitational weight. The result is a colossal thermonuclear explosion.

What could happen if a supernova explosion occurred relatively close to Earth—that is, "within a few tens of light years from us"?[42] Although ultraviolet (UV) light, X-rays, and gamma rays would bombard the planet, Earth's atmosphere would deflect this incoming radiation. The danger lies in the possibility that a supernova's *cosmic rays* destroy the protective layer of ozone around the planet, thus enabling "the penetration of UV radiation and absorption of visible sunlight by NO_2 in the atmosphere."[43] UV radiation can cause skin cancer, while NO_2 can form ground ozone, an air pollutant, when it mixes with oxygen in the presence of bright light.

(b) *Gamma-ray bursts.* There are two types of gamma-ray bursts: long and short. The former last for more than two seconds, constitute 70 percent of observed gamma-ray bursts, and probably arise from supernova explosions. In contrast, the latter last for less than two seconds and their cause is not entirely known to science. The primary danger posed by gamma-ray bursts of both types is their potential to "destroy the ozone layer and create enormous shocks going through the atmosphere, provoke giant global storms, and ignite huge fires."[44] The amount of UV radiation could also exceed the lethal dose for humans. Furthermore, as the cosmologist Arnon Dar notes, "the short duration of [gamma-ray bursts] and the lack of an [early] warning signal . . . make protection by moving into the shade or a shelter, or by covering up quickly, unrealistic for most species."[45]

(c) *Entropy death*. Another feature of our cosmic risk background not previously noted is the entropy death of the cosmos, a topic that falls within the scientific field of *physical eschatology*. For the sake of context, let's start at the beginning: about 13.8 billion years ago, the universe began with a "big bang." This term does not refer to an explosion but the *moment* at which the universe started expanding—a phenomenon that continues today at an accelerating rate. One might picture the universe growing within some larger space that envelops it, but this would be wrong: what is expanding is *space itself*.[46] By analogy, consider ants crawling on the surface of a balloon as it is being inflated. Just like the universe, this surface has *no center and no boundaries*. Furthermore, the bigger the balloon gets, the further away the ants find themselves, just as we observe clusters of galaxies drifting in opposite directions (resulting in the famous "redshift" phenomenon). As this expansion continues into the deep future, any trace of other galaxy clusters will eventually slip out of view forever.

The aging universe will become increasingly "chaotic" due to the inexorable rise of entropy, a measure of "the unavailability of a system's thermal energy for conversion into mechanical work, often interpreted as the degree of disorder or randomness in the system."[47] Our sun will eventually turn into a red giant, completely destroying Earth about 7.59 billion years from now—although it will become uninhabitable long before that.[48] Roughly 10^{40} years in the future, nearly all the protons in the universe will have decayed, making life of any sort impossible. The only entities occupying the universe at this point will be black holes, which will drift about for another 10^{30} years, until their mass converts into radiation via the quantum mechanical process known as "Hawking radiation." Finally, approaching 10^{100} years into the future, the universe will sink into a Great Dark Era during which the "available energy is limited and the expanses of time are staggering."[49]

But as the theoretical astrophysicist Fred Adams notes, it could be that the universe, at this late stage in its life cycle, undergoes a vacuum transition that introduces new laws of physics (see section 3.3). These laws could, speculatively speaking, give "the universe a chance for a fresh start."[50] Other scientists, such as Michio Kaku, have wondered about the possibility of our descendants (a) escaping into a nearby universe (within the multiverse) through a wormhole, (b) migrating into a baby universe of our own creation, or (c) copying our universe and then transferring it into another universe as a "seed."[51] While these ideas are hardly more than wild conjectures in the dark, they offer a faint glimmer of hope that our distant posthuman progeny could overcome the *cosmic nihilism* implied by the entropy death of the universe.

In a phrase, the great epitaph of the cosmos is "In with a bang, out with a whimper."[52]

Chapter 3: Unintended Consequences

3.1 Intended Causes, Unintended Effects

One of the most influential articles about unintended consequences was published in 1936 by the sociologist Robert Merton. It begins by noting that "in some one of its numerous forms, the problem of the unanticipated consequences of purposive action has been treated by virtually every substantial contributor to the long history of social thought." Yet the "diversity of context and variety of terms by which this problem has been known . . . have tended to obscure the definite continuity in its consideration."[1] According to Merton, there are five causes of unintended consequences, namely, (1) ignorance, (2) error, (3) pursuing immediate rather than longer-term interests, (4) value-constraints resulting in unfortunate outcomes, and (5) what can be called "self-defeating prophecies," or predictions that prevent the predicted event from taking place.[2]

What is an unintended consequence? Merton doesn't provide a succinct definition in his paper, but the theorist Langdon Winner does in his 1977 book *Autonomous Technology*. According to Winner, unintended consequences "are almost always negative or undesirable effects" that "are not *not* intended." The double negative here "means that there is seldom anything in the original plan that aimed at preventing them."[3] One could interpret this as distinguishing between (a) purposive actions that aim to avoid an undesirable outcome but fail, and (b) purposive actions that cause an outcome that no one even hoped wouldn't happen. Some of the phenomena discussed below are of the latter sort, but others involve people or governments simply

75

failing to take the necessary precautions to avoid a disaster.

While the history of unintended consequences is as long as the history of purposive action itself, climate change and biodiversity loss are the first unintended consequences with existential effects. But given that the human population is growing and the power of technology is increasing, we should expect more unintended consequences with survival implications in the future. The following sections examine four types of unintended consequences that could threaten the perpetuation of our species.

3.2 Climate Change and Biodiversity Loss

Not since cyanobacteria has a single taxonomic group been so in charge. Humans have proven we are capable of seismic influence, of depleting the ozone layer, of changing the biology of every continent.
—Jennifer Jacquet[4]

The term "climate change," as used in contemporary discussion, refers to the observed alterations of global weather patterns due to humans burning fossil fuels, a process that converts decomposed biological matter into usable energy while releasing carbon dioxide as a byproduct.[5] Here is what we know: carbon dioxide is transparent to visible light but not to infrared light. (Humans cannot see infrared light, but we can detect it as heat via the sensory modality of thermoception.) This property of carbon dioxide is notable because most of the electromagnetic radiation from the sun takes the form of visible light. Consequently, incoming light passes through the atmosphere, reaches Earth's surface, and is then reradiated as infrared light. Much of this reradiated light is unable to escape back into space because of (in part) carbon dioxide—a *greenhouse effect* that results in global warming and, thus, climate change.

Although carbon dioxide is the primary driver of climate change today, it isn't the only **greenhouse gas** (GHG) contributing to the problem. As Earth's surface heats up, permafrost in the northern regions of the globe is beginning to thaw. This contains frozen organic matter that will release dangerous amounts of *methane*, a GHG about 25 times more potent in trapping heat than carbon dioxide. The result could be a *positive feedback loop* such that the warmer Earth is, the warmer Earth becomes, and so on, until all the permafrost has melted. Another positive feedback loop involves the melting of glaciers and snow at the poles. As readers may know, the achromatic "color" of white contains all the wavelengths of visible light—i.e., white objects are those that reflect every color frequency by absorbing none. Thus, when visible light from the sun comes into contact with regions of Earth covered in snow, this radiation is reflected back into space *without* being converted into surface-warming infrared heat. As the snowcaps melt, though, more incoming light will be absorbed and reradiated as heat, thus amplifying the effects of global warming.

The existence of positive feedback loops in the climate system is worrisome because one or more could, in principle, initiate a **runaway greenhouse effect**.[6] According to a recent study, an atmosphere rich in water vapor—also a GHG—"absorbs more sunlight and lets out less heat than previously thought, enough to put the earth into a spiral from which there would be no return."[7] Something similar very likely happened on our planetary neighbor Venus, which succumbed to a water vapor–driven runaway greenhouse effect early in its history. Fortunately, this same study suggests that we would need to reach 30,000 parts per million (ppm) of carbon dioxide in the atmosphere to surpass a tipping point, and even the most pessimistic projections from the Intergovernmental Panel on Climate Change (IPCC) anticipate carbon dioxide levels of no more than about 1,000 ppm. (For historical levels, see Figure C.) But as an article in the *MIT Technology Review* notes, "there is an important caveat. Atmospheric physics is so complex that climate scientists have only a rudimentary understanding of how it works."[8] Thus, there could be unknown phenomena that render a runaway catastrophe far more probable than we currently believe it is. Uncertainty should foster an attitude of caution—a judicious maxim that applies to many scenarios in this book.

We have so far discussed the causes of climate change from a physics and chemistry perspective. But there are also social factors behind this phenomenon. Indeed, the ongoing release of huge quanti-

Figure C. Fluctuations of Earth's Atmospheric Carbon Dioxide Going Back 400,000 Years

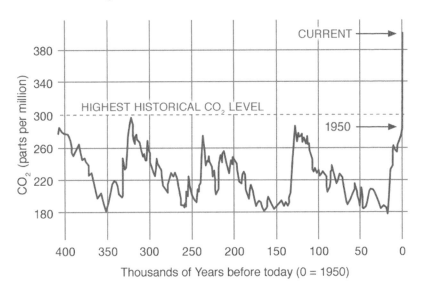

Source: U.S. National Oceanic and Atmospheric Administration

ties of carbon dioxide—more than 9 gigatons per year since 2010—stems from what game theorists call a "cooperation problem," specifically the **tragedy of the commons**.[9] To quote a canonical 1968 article by Garrett Hardin at length:

> Picture a pasture open to all. It is to be expected that each herdsman will try to keep as many cattle as possible on the commons. Such an arrangement may work reasonably satisfactorily for centuries because tribal wars, poaching, and disease keep the numbers of both man and beast well below the carrying capacity of the land. Finally, however, comes the day of reckoning, that is, the day when the long-desired goal of social stability becomes a reality. At this point, the inherent logic of the commons remorselessly generates tragedy.
>
> As a rational being, each herdsman seeks to maximize his gain. Explicitly or implicitly, more or less consciously, he asks,

"What is the utility to me of adding one more animal to my herd?" This utility has one negative and one positive component.

1) *The positive component is a function of the increment of one animal. Since the herdsman receives all the proceeds from the sale of the additional animal, the positive utility is nearly +1.*

2) *The negative component is a function of the additional overgrazing created by one more animal. Since, however, the effects of overgrazing are shared by all the herdsmen, the negative utility for any particular decision-making herdsman is only a fraction of -1.*

Adding together the component partial utilities, the rational herdsman concludes that the only sensible course for him to pursue is to add another animal to his herd. And another; and another. . . . But this is the conclusion reached by each and every rational herdsman sharing a commons. Therein is the tragedy. Each man is locked into a system that compels him to increase his herd without limit—in a world that is limited. Ruin is the destination toward which all men rush, each pursuing his own best interest in a society that believes in the freedom of the commons. Freedom in a commons brings ruin to all.[10]

Unfortunately, the commons tragedy is even more insidious in the case of climate change because of how many people are involved. Seven and a half billion polluters and counting makes the discrete contributions of each person to the problem seemingly negligible, thus yielding a weak sense of moral responsibility for the aggregate outcome.[11] Two related issues are noted by Ingmar Persson and Julian Savulescu:

If the number of agents involved is large, it also becomes harder to establish the trust necessary for cooperation because the individual agents are unlikely to know each other. So, it is unlikely that they will have developed concern and liking for each other. Likewise, it will be harder for them to keep an eye on each other and check whether there is free-riding.[12]

According to the best current science, the consequences of climate change will be "severe," "pervasive," and "irreversible."[13] As a 2016 paper in *Nature* by more than 20 scientists from around the world explains, the fossil fuels that humanity is burning right now will have effects lasting for some 10,000 years, perhaps affecting more human beings than have ever before existed.[14] As the authors put it,

The next few decades offer a brief window of opportunity to minimize large-scale and potentially catastrophic climate change that will extend longer than the entire history of human civilization thus far. Policy decisions made during this window are likely to result in changes to Earth's climate system measured in millennia rather than human lifespans, with associated socioeconomic and ecological impacts that will exacerbate the risks and damages to society and ecosystems that are projected for the twenty-first century and propagate into the future for many thousands of years.[15]

Among the specific consequences of climate change are extreme weather events, megadroughts lasting decades, severe coastal flooding, sea-level rise (partly due to thermal expansion), melting glaciers and the polar icecaps, desertification, deforestation, food supply disruptions, natural epidemics and pandemics, and a host of societal quandaries like mass migrations, social upheaval, economic collapse, and political instability. Studies have also found that lightning strikes will increase by 50 percent by 2100, allergy seasons will become "longer and more intense," and Earth's tilt and rotational speed will change.[16] One study even suggests that certain regions of the globe will experience heat waves that surpass the 95 degree "wet bulb" threshold beyond which our natural thermoregulatory mechanisms are no longer effective.[17] In other words, no human would survive such weather even if she were standing naked in the shade next to a giant fan.

Historically speaking, the hottest 17 years on record have all occurred since 2000, that is, with the single exception of 1998. Beginning with the hottest, the record holders are 2016, 2015, 2014, 2010,
2013, 2005, 1998, 2009, 2012, 2003, 2006, 2007, 2002, 2004, 2011,
2001, and 2008.[18] While 2017 is unlikely to break 2016's record, the
U.K.'s Met Office projects that it "will still rank among the hottest
years on record."[19]

* * *

Yet another consequence of climate change not listed above is *global
biodiversity loss*. The term "global biodiversity" refers to the overall
variety of living creatures on the planet, including "all organisms, species, and populations; the genetic variation among these; and their
complex assemblages of communities and ecosystems."[20] Additional
factors driving biodiversity loss are habitat destruction, ecosystem
fragmentation, invasive species, overexploitation, and pollution—all
of which are the result of human activity. Although biodiversity loss
has received considerably less attention from the popular media than
climate change, it could pose a threat to human prosperity and survival that is nearly as dire. Indeed, the curtailing of biological diversity
will very likely be humanity's greatest legacy on this pale blue dot (unless we do something like convert the planet into computronium).[21]

To illustrate the seriousness of this problem, consider the declining populations of wild species around the world. According to the
third Global Biodiversity Outlook (GBO-3) report from 2010, the total population of wild vertebrates within the tropics—that is, between
the Tropic of Cancer and the Tropic of Capricorn—fell by an incredible 59 percent in only 36 years, from 1970 to 2006. (The taxon of
vertebrates includes mammals, birds, fish, reptiles, and amphibians.)
The report also found that vertebrates in freshwater environments declined by 41 percent, farmland birds in Europe declined by 50 percent
since 1980, birds in North America declined by 40 percent between
1968 and 2003, and about 25 percent of all plant species—the foundation of the food chain—are currently "threatened with extinction."[22]

Four years after the GBO-3, the World Wide Fund for Nature

published the 2014 Living Planet Report. This found that between 1970 and 2010, the global abundance of wild vertebrates dropped by a staggering 52 percent. The 2016 Living Planet Report updated this number, finding that wild vertebrates declined by 58 percent between 1970 and 2012. Although these studies don't extrapolate such trends into the future, readers are welcome to do so, noting that as ecosystems weaken the likelihood of further population losses will tend to increase. In fact, a 2006 study published in *Science* projects observed patterns of marine biodiversity loss into the twenty-first century, concluding that unless significant changes are made to human behavior, there will be virtually *no more wild-caught seafood by 2048*.

The 2016 Living Planet Report leaves us a grim observation and a warning: (i) it calculates that "by 2012, the biocapacity equivalent of 1.6 Earths was needed to provide the natural resources and services humanity consumed in that year."[23] And (ii) it cautions that, unless immediate action is taken to avert a disaster, "we could witness a two-thirds decline [of vertebrates] in the half-century from 1970 to 2020."[24]

Other studies confirm the general findings of these reports. For example, about 20 percent of all reptile species, 50 percent of freshwater turtles, and 60 percent of the world's primates are currently under threat.[25] There has also been a disconcerting decline in "the most important insect that transfers pollen between flowers and between plants," namely, the honey bee. This phenomenon is called *colony collapse disorder*, and it could have major implications for agricultural production in the future.[26]

As for the diversity of marine life, increasing concentrations of carbon dioxide in the atmosphere are causing the world's oceans to acidify. One consequence is a phenomenon called "coral bleaching," whereby coral lose their zooxanthellae, a type of algae that they need to survive. About 10 percent of coral reefs today are underwater ghost towns, and some 60 percent are in danger of bleaching. This has direct consequences for humanity because coral reefs "provide us with food, construction materials (limestone) and new medicines," and "more than half of new cancer drug research is focused on marine organisms."[27] Yet another scientific study found that the *rate* of ocean acidification today is comparable to the changes that occurred during the Permian-Triassic mass extinction, also dubbed the "Great Dying," during which some 95 percent of all species died out.[28] As the science journalist Eric Hand writes, "The [Permian] extinction holds a cautionary lesson for today: Because of CO2 released by burning fossil fuels, oceans could now be acidifying *even faster* than they did 250 million years ago, although the process hasn't yet persisted nearly as long."[29] He adds that

> *The Permian-Triassic catastrophe holds mixed messages for Earth today. On the one hand, the pace of acidification was slower than it is now. The study team estimates that, in the acidification event, 24,000 gigatons of carbon were injected into the atmosphere over 10,000 years—a rate of 2.4 gigatons per year—and most of it wound up in the oceans. Currently, scientists estimate carbon from all sources is entering the atmosphere at a rate of about 10 gigatons per year.*
>
> *On the other hand, today's economically viable fossil fuel reserves contain only about 3,000 gigatons of carbon—far shy of the Permian total, even if human beings burn it all.[30]*

So "we're injecting the carbon faster, but it's unlikely that we have as much carbon to inject"—although this does not offer much reassurance, since the evolutionary mechanisms that ensure a sufficient degree of organismal adaptedness operate on slow, even geological timescales.[31] As Rachel Wood, a coauthor of the study, puts it, "The data is compelling and we *really should* be worried in term[s] of what is happening today."[32]

Human pollution is also carving out large regions of the ocean in which the amount of dissolved oxygen is too low for organisms to survive. These regions are called **dead zones**, and the most recent count by the marine biologist Robert Diaz and his colleagues found more than 500 around the world.[33] The biggest dead zone discovered so far is located in the Baltic Sea, and scientists estimate it to be about 27,000 square miles, or a little smaller than New Hampshire, Vermont, and Maryland put together. Furthermore, recent ocean expeditions have discovered "islands" of plastic trash in the gyres (i.e., the large circulating currents) of the Indian Ocean, Atlantic Ocean, and Pacific Ocean. The trash heap of the last is called the "Great Pacific Garbage Patch," and scientists estimate it to be up to "twice the size of the continental United States."[34] Sadly, these *waterfills* are likely to grow even larger as consumerist societies become bloated and more numerous; some researchers estimate that there will be, by weight, more plastic than fish in the world's oceans by 2050.

The devastation caused by human activity, though, has not merely resulted in declining *populations* of species around the world but a marked loss of the *total number* of species.[35] For example, the extinction rate today is thought to be between 100 and 1,000 times higher than the natural "background extinction rate" that operates during normal periods of biological history. As a 2015 study in *Science Advances* puts it, there has been "an exceptionally rapid loss of biodiversity over the last few centuries, indicating that a sixth mass extinction is already under way."[36] Let's call this event the *Anthropocene extinction*, where the "Anthropocene" is a proposed geological epoch that began around the 1950s.[37] Only five mass extinction events have previously occurred in the 3.8 billion-year duration of life on Earth; these are known as the Big Five. Thus, the evidence now suggests that we should talk about the **Big Six**. (This is, indeed, why biodiversity loss could be our longest-lasting legacy on the planet.)[38]

Finally, as a result of these evolutionarily rapid disruptions to the biosphere, we may be approaching an ecological threshold that, if crossed, would initiate a sudden, irreversible, catastrophic collapse of the global ecosystem. As the authors of a 2012 study published in *Nature* write, a planetary-scale transition or "state shift" could precipitate "substantial losses of ecosystem services required to sustain the human population." (An *ecosystem service* is any ecological process that benefits humanity, such as food production and crop pollination.) If this were to occur, it could cause "widespread social unrest, economic instability, and loss of human life."[39] One of the paper's coauthors, the ecologist Adam Smith, notes that this could happen in a matter of decades—that is, within a single human's lifetime, and perhaps within the lifetimes of people alive today.[40] (See Box 4.)

Box 4. An important concept that ties many of these phenomena together is that of a **planetary boundary**. According to a 2009 report authored by nearly 30 scientists from around the world, including several Nobel laureates, there are nine Earth-system processes associated with planetary boundaries, namely: (1) climate change, (2) ocean acidification, (3) stratospheric ozone depletion, (4) atmospheric aerosol loading, (5) biogeochemical flows (i.e., phosphorus and nitrogen cycles), (6) global freshwater use, (7) land-system change, (8) rate of biodiversity loss, and (9) chemical pollution. Together, these outline a "safe operating space for humanity" in which sustainable development must proceed or else risk disaster. As the authors write, "Anthropogenic pressures on the Earth System have reached a scale where abrupt global environmental change can no longer be excluded. . . . Transgressing one or more planetary boundaries may be deleterious or even catastrophic due to the risk of crossing thresholds that will trigger non-linear, abrupt environmental change within continental- to planetary-scale systems." The authors add that the "proposed boundaries are rough, first estimates only, surrounded by large uncertainties and knowledge gaps," a point that should make us *more* rather than *less* wary. Still, they write that the planetary boundaries concept "lays the groundwork for shifting our approach to governance and management . . . toward the estimation of the safe space for human development. Planetary boundaries define, as it were, the boundaries of the 'planetary playing field' for humanity if we want to be sure of avoiding major human-induced environmental change on a global scale." Unfortunately, it appears that "humanity has already transgressed three planetary boundaries: for climate change, rate of biodiversity loss, and changes to the global nitrogen cycle," meaning that we have created a situation in which we are vulnerable to global environmental shifts that could unfold rapidly and severely affect the stability and perpetuation of human civilization.*

* Rockström, Johan, et al. 2009. Planetary Boundaries: Exploring the Safe Operating Space for Humanity. *Ecology and Society*. 14(2). URL: https://www.ecologyandsociety.org/vol14/iss2/art32/main.html.

There is, to be sure, a crucial difference between the two attitudes of *being alarmed* and *being an alarmist*: the former is warranted by the evidence (on the totality condition) whereas the latter is not. Given the brief survey of environmental data above, humans in the twenty-first century have every reason to be alarmed—this is, in fact, why some environmentalists have gone so far as to advocate criminalizing climate denial propaganda.[41] Not only could climate change and biodiversity loss push civilization to the brink of collapse, but as we will explore in subsection 4.3.3, the *conflict-multiplying effects* of these context risks will have major exacerbatory consequences for certain agential risks—sequelae that could nontrivially increase the overall likelihood of an existential catastrophe.

Before moving on to the next section, we should consider a few additional consequences of climate change and biodiversity loss that are relevant to existential risk studies. First, research shows that the degree to which people discount the future is, in part, a function of "the stability or instability of their environment."[42] In other words, highly unstable environments lead people to discount the future more, whereas stable environments lead to less steep discount rates. As one scholar puts it, "It doesn't pay to save for tomorrow if tomorrow will never come, or if your world is so chaotic that you have no confidence you would get your savings back."[43] We should thus expect that climate change and biodiversity loss will decrease interest in existential risk studies in the coming decades, given that societies will likely be preoccupied with more immediate concerns and less confident about their long-term prospects. The positive correlation between environmental instability and steep discounting could be bad news for existential risk scholarship.[44]

Second, studies show that carbon dioxide concentrations can have appreciable negative effects on cognition. One study, for example, found "moderate" declines in cognitive performance on decision tasks when the concentration of carbon dioxide was increased from 600 to 1,000 ppm, and an "astonishingly large" drop in performance from 1,000 to 2,500 ppm.[45] This is worrisome because our species has spent nearly all its evolutionary history breathing in air with a carbon dioxide concentration of between 180 and 280 ppm. As a result of industrialization, though, carbon dioxide concentrations have risen dramatically. In fact, we recently passed the disheartening milestone of 400 ppm of carbon dioxide in the atmosphere, which is irreversible in the foreseeable future, and some research suggests that carbon dioxide levels could reach 1,000 ppm in the ambient air by the end of this century.[46] Consequently, there could be widespread cumulative effects on our capacity to solve the increasingly complex problems before us—or, put differently, subtle losses of intellectual capacity due to carbon dioxide emissions could non-negligibly increase the overall risk by compromising our collective intelligence. "In effect," the journalist Daniel Grossman writes, "the fuel we burn might not only warm the planet but could also make us a bit dumber."[47]

Environmental degradation poses monumental challenges for humanity. But it is not the only unintended disaster with existential implications.

3.3 Physics Disasters

Particle accelerators smash beams of particles traveling at close to the cosmic speed limit of light in order to study the elementary constituents of the physical universe. The biggest and most powerful particle accelerator in the world today is the Large Hadron Collider (LHC), built by the European Organization for Nuclear Research (CERN) beneath the Franco-Swiss border near Geneva. The LHC started running in 2008, and it boasts of having discovered the Higgs boson—a previously elusive particle in the Standard Model of particle physics—between 2012 and 2013. But there are a number of possible risks associated with focusing huge quantities of energy in tiny regions of space. Three potential hazards are especially notable, namely, the formation of a black hole, the transformation of the matter out of which our planet is made into "strange matter," and the creation of a "true vacuum."[48] Let's examine these in turn:

(a) *Black holes.* Current theory suggests that "it is impossible for microscopic black holes to be produced at the LHC." Nonetheless, some more speculative theories predict that the LHC could produce them, although the same theories predict that they would decay immediately. Even if they didn't, though, "hypothetical stable black holes can be shown to be harmless." This is because black holes of this sort can also result from naturally occurring cosmic rays, which nature "routinely produces." Thus, the LHC Safety Report argues that "the fact that the Earth and Sun are still here rules out the possibility that cosmic rays or the LHC could produce dangerous charged microscopic black holes."[49]

(b) *Strangelets.* A strangelet is a hypothetical clump of *strange matter.* Scientists have raised the possibility that strangelets could form in particle accelerators and "coalesce with ordinary matter and change it to strange matter." But this appears to be unlikely because "it is difficult for strange matter to stick together in the high temperatures produced by [particle] colliders, rather as ice does not form in hot water."[50]

(c) *Vacuum bubble.* Scientists have also speculated that the universe might not be in its most stable state and consequently that physical disturbances could tip it into a more stable state. If this were to happen, the result would be "a bubble of 'true vacuum' expanding outwards at the speed of light, converting the universe into [a] different state apparently inhospitable for any kind of life."[51] In an influential paper, Piet Hut and Martin Rees argue that the probability of this occurring "is completely negligible since the region inside our past light cone has already survived some 105 cosmic ray collisions" that could induce a vacuum bubble.[52] Yet the earth, moon, and observable stars still exist, which suggests that the danger is minuscule.

There are a number of criticisms to be made of the reassuring conclusions reached by LHC safety evaluators and other scientists. First, the *cosmic ray argument* used by Hut and Rees to argue that particle accelerator experiments are safe fails to take into account the observation selection effect (discussed in section 1.6). This states that history cannot provide useful information about the riskiness of certain phenomena if these phenomena are incompatible with the existence of observers like us. Whether a vacuum bubble catastrophe, for example, is probable or improbable, all observers should expect to see an intact universe around them, precisely as we do. Nick Bostrom and the cosmologist Max Tegmark point this out in a 2005 article, showing that the safety assessments of at least one major particle accelerator—which ran for five years before Bostrom and Tegmark's paper—relied on flawed arguments.[53]

Another problem is that scientific evaluations of the threats posed by particle accelerators are conditioned upon the given arguments being sound—a rarely acknowledged but crucial point. Thus, one must consider not only the probability assigned to a risk scenario *by* a given study but also the probability that the study *itself* is flawless. In a coauthored article, the scholars Toby Ord, Rafaela Hillerbrand, and Anders Sandberg identify three types of errors that could affect the outcomes of probability estimates. The first two are well-known: there could be model or parameter errors. A model is a picture of reality derived from a theory. If the underlying theory is wrong—meaning incorrect or incomplete—then the model may contain a flaw. Alternatively, if the inputs for the model are incorrect, then so will the outputs, even if the model is accurate. Third, they argue that, independent of these two phenomena, there could be *calculation mistakes* that yield faulty outcomes. For example, the Mars Climate Orbiter spacecraft failed not because of any model or parameter error but "because a piece of control software from Lockheed Martin used Imperial units instead of the metric units the interfacing NASA software expected." As the authors note, "Calculation errors are distressingly common."[54] This provides yet another reason to be skeptical—albeit tentatively, as all skepticism should be—of the conclusions of risk assessments, especially when the relevant consequences could be cosmically catastrophic.

Finally, there could be existential risks associated with certain high-powered physics experiments that require a theory X to specify, but understanding theory X requires a series of concepts A, B, and C that lie outside our **cognitive space**.[55] That is to say, it could be that the mechanisms in our brains responsible for generating concepts are simply not up to the task of generating A, B, or C. Without

Box 5. To be more explicit, the idea behind the cognitive space thesis is this: (a) concepts are mental representations of the world, meaning that one can only represent—and therefore know about—aspects of reality if one can grasp the relevant concepts; (b) the concept-generating mechanisms in our brains are intrinsically limited with respect to their output as a result of evolution; it follows that (c) there may be concepts that fall outside of our cognitive space, and consequently facts about reality that are not merely unknown, but unknowable *in principle* rather than *in practice*. This is an issue that we will return to again in subsequent chapters.

these concepts, we are unable to represent the corresponding parts of reality, which in this case pose grave dangers to human existence (see Box 5). This isn't an issue of human science not discovering the relevant theory in time but of human science never being able to reach it in principle. By analogy, the canine brain is simply unable to comprehend the risks associated with nuclear weapons. No matter how well-trained or clever, no dog will ever grasp the concept of a *nuclear chain reaction*—just as, to draw a perceptual analogy, no matter how good one's vision the human eye will never see light in the microwave frequency band. So, we could be in a similar epistemic predicament with respect to any number of concepts that, if only we were to grasp them and, therefore, emerge from our Platonic cave, would lead us to exclaim, "Stop the physics experiment immediately!"

All things considered, the risks posed by particle accelerators do appear small. But there could be parameter, model, or calculation errors as well as fundamental limitations to our cognitive space that prevent us from accurately assessing the true threat level.

3.4 Geoengineering

As section 3.2 established, the effects of climate change will be severe and pervasive. Consequently, a government, group, or wealthy individual could opt to modify the climate through one or more processes of *geoengineering.* This comes in two general varieties: (1) carbon dioxide removal (CDR), and (2) solar radiation management (SRM). The first focuses on the primary chemical cause of climate change, namely, the accumulation of carbon dioxide in the atmosphere. CDR techniques include (a) carbon sequestration, which encompasses a wide range of techniques, such as removing carbon dioxide from the flue gases released by power stations and storing it, and (b) ocean fertilization, which entails "fertilizing" the oceans by introducing nutrients like iron or nitrogen to encourage the growth of phytoplankton, thereby increasing the mass of carbon sinks.

While CDR could pose a number of risks—e.g., by disrupting marine ecosystems that are already under stress—the greatest dangers stem from its cousin, SRM. This approach does not attempt to change the amount of atmospheric carbon dioxide but rather focuses on the other major ingredient of climate change: the incoming electromagnetic radiation from the sun. Options include placing giant "space mirrors" in the inner Lagrangian point, L1, between Earth and the sun to redirect incoming light, and making the track of waves created by cargo ships on the ocean "brighter" and more "foamy" to enhance Earth's albedo—an idea called "wake whitening."[56] The most-discussed option, though, involves injecting a sulfate aerosol such as hydrogen sulfide, sulfur dioxide, or sulfuric acid into the stratosphere. This could be accomplished using balloon-borne probes, aircraft, or even "factories on the ground [that] pump sulfur dioxide upward through miles-long hoses, their nozzles held aloft by high-flying zeppelins."[57] The stratospheric particles would then reflect incoming solar radiation back into space—as occurs after a large volcanic eruption—thereby inducing an effect known as *global dimming* to counteract global warming.

But modifying stratospheric chemistry could have many unforeseen negative consequences, perhaps producing effects that are even worse than unchecked global warming. No one knows for sure what would happen, given the complex and chaotic nature of the climate system. As Lawrence Krauss puts it, "At this point . . . the unknowns outweigh the knowns."[58] There is also the risk of a *double catastrophe scenario* whereby an ongoing SRM regime—that is, an SRM regime that has already been implemented—is suddenly terminated due to unpredictable vagaries like war, terrorist attacks, shifts of political power, economic recessions, and so on. If this were to occur, it could produce global agricultural failures that cause widespread famines or maybe even a runaway greenhouse effect that renders Earth uninhabitable. The scholars Seth Baum, Timothy Maher, and Jacob Haqq-Misra call this danger the problem of "intermittency": once a SRM regime is established, it *must not be interrupted*, especially not abruptly. Yet an abrupt interruption cannot be ruled out—and thus the danger.[59]

Moving forward, as the environmental situation becomes more dire, stratospheric geoengineering may become quite attractive to actors looking for an easy "technological fix" for climate change. For example, it would be relatively cheap to implement a SRM regime—far cheaper than mitigating climate change directly (see section 6.5)—and this could entice states, groups, or individuals to act unilaterally. The problem is that, to quote a Global Challenges Foundation report, "individual states acting alone may be less likely to properly take into account the interests of other states and may [not] be concerned about catastrophic consequences in other regions."[60] Complicating matters even further, individual states need not be motivated by *selfish impulses* to ensure their own survival; they could instead be inspired by *altruistic concerns* for the well-being of humanity as a whole. But if one state were to act alone, the outcome could very well be bad for everyone, a scenario that Nick Bostrom, Anders Sandberg, and Tom Douglas call the **unilateralist's curse**. In their words, "Let a unilateralist situation be one in which each member of a group of agents can undertake or spoil an initiative regardless of the cooperation or opposition of other members of the group."[61] They elaborate this idea as follows:

> *Each agent decides whether or not to undertake X on the basis of her own independent judgment of the value of X, where the value of X is assumed to be independent of who undertakes X, and is supposed to be determined by the contribution of X to the common good. Each agent's judgment is subject to*

error—some agents might overestimate the value of X, others might underestimate it. If the true value of X is negative, then the larger the number of agents, the greater the chances that at least one agent will overestimate X sufficiently to make the value of X seem positive. Thus, if agents act unilaterally, the initiative is too likely to be undertaken, and if such scenarios repeat, an excessively large number of initiatives are likely to be undertaken.[62]

Bostrom, Sandberg, and Douglas argue that lifting this curse requires universal adherence to a *principle of conformity*, which states that "when acting out of concern for the common good in a unilateralist situation, reduce your likelihood of unilaterally undertaking or spoiling the initiative to a level that *ex ante* would be expected to lift the curse." This can be accomplished through what the authors call "collective deliberation," "epistemic deference," and "moral deference."[63] The first involves sharing "data and reasoning between agents in the hope that this will resolve their disagreement about the desirability of proceeding with the contested initiative." The second aims "to appeal to each agent's reflective rationality," thus prompting all agents in "an epistemic disagreement [to] reflect on the fallibility of their own judgment and adjust their posterior probability to take into account the fact that other agents have different opinions." The third states that if the first two options fail, "it might nevertheless be possible for the group to lift the curse if each agent complies with a moral norm which reduces the likelihood that he acts unilaterally, for example, by assigning decision-making authority to the group as a whole or to one individual within it."[64]

With respect to both CDR and SRM, state and nonstate actors alike should take seriously the principle of conformity, however dire the climatic situation becomes. We will return to this topic in section 6.5.

Chapter 4: Agent-Tool Couplings

4.1 Conceptual Framework

Although the phenomena above are extremely worrisome, most existential risk scholars concur that the greatest threats to humanity stem from the misuse and abuse of advanced technologies. These technologies are historically recent—the first nuclear weapons being developed less than eight decades ago—so humanity has no track record of surviving the associated risks. Indeed, some have argued that extraterrestrial civilizations at our level of scientific development tend to destroy themselves, and that this explains the disquieting Great Silence of section 1.5.[1] There may, in fact, be something to this line of reasoning: metaphorically speaking, one could describe contemporary humanity as a pyromaniacal child whose matches have suddenly been replaced by a flamethrower capable of burning down the entire global village with one pull of the trigger. What are the chances that a child who plays with fire will survive with this new weapon in her hands? What is the likelihood that she will avoid self-immolation?[2]

So far, most research on agent-tool risks has focused on the relevant tools, which we previously called "WTDs," for "weapons of total destruction." Studying WTDs is important because understanding their unique properties could enable risk scholars to devise more effective risk mitigation strategies. More recent work, though, has concentrated on the different types of agents who, motivated by normative ideologies and subject to error, might initiate a WTD catastrophe. This phenomenon is no less important to study than advanced technologies because the risks posed by agent-tool couplings depend on the properties of *both* the tools and the agents.[3]

To emphasize this point, consider the *two worlds thought experiment*: say that a world X is cluttered with universally accessible WTDs, whereas a world Y contains only a single WTD. Now ask: In which world would a rational person prefer to live? The best answer appears to be world Y, since it contains fewer WTDs. But deciding to inhabit world Y based on this information alone would be unwise. One should also consider information about the agents who populate each world. Thus, imagine further that world X is populated by a single species of pacifistic peaceniks, whereas world Y is run by genocidal warmongers. With this additional information we can ask the same question as before: in which would a rational person prefer to reside? The best answer now appears to be world X, despite its oversized arsenal of WTDs. This "experiment" merely underlines that the overall riskiness of a world is the product of multiple factors, includ-

---

Box 6. One should not interpret the agent-tool framework as endorsing what philosophers of technology call the **neutrality thesis**. According to this thesis, technologies are intrinsically neutral objects—that is, "mere tools"—that do not influence how agents use them or behave. Many philosophers reject this thesis, arguing instead that technologies are crucially shaped by the values of their designers and can, in turn, crucially shape the attitudes and actions of their users; as the media theorist Marshall McLuhan puts it, "We shape our tools and afterwards our tools shape us." For example, a gun could serve multiple functions, such as propping open a door or digging a hole in one's garden. But its particular design makes it far more suitable for a narrow range of specific functions, such as shooting a home invader or robbing a bank. The phrase "Guns don't kill people, people kill people" thus fails to account for how the gun side of the *person-gun coupling* can reconfigure the inclinations, capabilities, and actions of the person side. The point is that talk of "tools" in this book shouldn't imply that the agents coupled to them are unaffected by the design features and values embedded in the corresponding artifacts.

ing its citizens, the existing technologies, and how these two interact.[4] (Indeed, see Box 6.)

We can further elucidate this conclusion with an idea already hinted at above: all advanced technologies have some *risk potential*, or capacity to inflict harm on humanity. But this risk potential can only be realized, under normal circumstances, by a complete agent-tool coupling. Thus, even though world X has (much) more risk potential, this is less likely to be realized given the irenic character of its residents, and *the probability of risk realization is ultimately what matters*. The opposite is the case for world Y, in which it seems to be only a matter of time before *someone* uses the only WTD there to obliterate the species.

The following sections will examine both sides of the agent-tool coupling, in reverse order.[5]

4.2 World-Destroying Technologies

> *Technology is giving life the potential to flourish like never before . . . or to self-destruct.*
> —*Future of Life Institute*

How could someone who wants to destroy the world accomplish this goal? What types of technologies could enable an agent to effectuate the collapse of civilization? Here we identify four categories of technologies that have the potential to be, or to enable, WTDs. Before examining these technologies, though, we will establish a few key concepts and techno-developmental trends that make emerging technologies—most notably biotechnology, synthetic biology, molecular nanotechnology, and "tool AI"—especially worrisome.

4.2.1 Dual Usability, Power, and Accessibility

The capacity of an agent to damage civilization is limited by the technological means at her or his disposal. For nearly our entire evolutionary history—some 200,000 years long—humanity has lacked the means to bring about a disaster of existential proportions. This is no longer the case today, of course, as there are enough nuclear weapons on the planet to "destroy the world many times over."[6] But it could be that by the end of the century, or perhaps within the next few decades, nuclear weapons will become the least of our troubles. The reason pertains to three specific phenomena, namely:

(i) Biotechnology, synthetic biology, nanotechnology, and tool AI are all **dual-use** in nature. This term originally referred to artifacts that have both military and civilian applications, but it has come to refer more generally to technologies, research, information, theories, and so on that can be employed for morally good or morally bad ends. For example, the very same knowledge of microbiology and genetics that could enable someone to discover a new cure for Ebola could also empower a terrorist to weaponize this virus. Similarly, the very same nanofactory (see below) that one could use to manufacture supercomputers for people in the developing world at virtually no cost could also be used to produce a warehouse full of exceptionally dangerous weaponry.

What is crucial to note about dual usability is that the good and bad properties cannot be separated: they are a package deal, and to eliminate either is to eliminate both.[7] In a phrase, emerging technologies are intrinsically and irremediably risky.

(ii) These technologies are also becoming increasingly powerful, thereby allowing people to manipulate and rearrange the physical world in ever more significant ways. Furthermore, this trend appears to be *exponential* (or even *exponentially exponential*) in multiple domains. For example, Moore's Law describes the doubling of transistors in an integrated circuit every two years, Rose's Law describes the exponential growth of "the number of qubits of quantum computers," and Butter's Law of Photonics states that "the amount of data one can transmit using optical fiber is doubling every nine months."[8] Other exponential trends pertain to computational capacity, electrical efficiency, rising product quality relative to falling price, and computer memory.

Similar growth rates can be found in biotechnology, synthetic biology, and nanotechnology—a general phenomenon captured by Ray Kurzweil's **Law of Accelerating Returns**.[9] For example, a 2014 *Nature* article observes that the cost of sequencing an average human genome has dropped at a rate that "does not just outpace Moore's law—it makes the once-powerful predictor of unbridled progress look downright sedate."[10] In a 2006 article, the *Economist* dubs this the "Carlson curve," after the researcher Rob Carlson, who drew "some graphs of the growing efficiency of DNA synthesis that . . . look suspiciously like the biological equivalent of Moore's law."[11] And an article by the founding chair of the U.S. National Science and Technology Council's subcommittee on nanotechnology, Mihail Roco, shows explosive growth rates with respect to how many nanotech research papers are being published, "the number of researchers and workers involved in one domain or another of nanotechnology," and "the value of products incorporating nanotechnology as the key component."[12]

So, the capacity of agents to reconfigure the world—for better or worse—is not only increasing, but increasing at an accelerating pace (see Figure D).[13]

Figure D. Growing Capacity of Agents to Kill

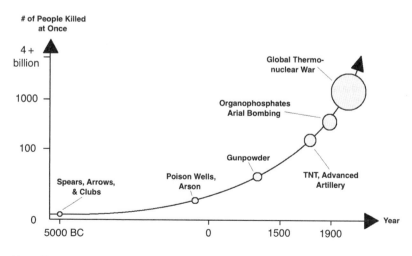

Note: Based on a graph created by Gary Ackerman

(iii) Lastly, advanced technologies are becoming more accessible to small groups and even individuals. We can identify four distinct axes along which this trend is unfolding, namely, intelligence (raw brainpower), knowledge (know-that), skills (know-how), and equipment (material means).[14] Taking these in order:

(a) The *cognitive abilities* needed to, for example, engineer microbes in a makeshift laboratory are less today than in the past. One does not need to be an "evil genius" to synthesize a designer pathogen or wreak havoc on society using some other emerging technology. The result is what the theorist Eliezer Yudkowsky dubs "Moore's Law of Mad Science," which asserts that "every eighteen months, the minimum IQ necessary to destroy the world drops by one point." However facetious this nomological generalization may appear (or was intended) to be, it captures something real and significant about the direction of technological development.

(b) The *amount of knowledge* necessary to exploit advanced technologies is declining. Information about all sorts of sensitive issues—including the full genomes of Ebola and smallpox, and possibly the blueprints for a nuclear bomb—are accessible online.[15] This touches upon an important issue that Bostrom calls "information hazards," or "risks that arise from the dissemination or the potential dissemination of true information that may cause harm or enable some agent to cause harm."[16] The Internet, in particular, has elevated hazards of this sort to new heights, enabling anyone with a smartphone to acquire dangerous strings of 1s and 0s. Even a book like this one could present an information hazard if it were to inspire the wrong people, although I have taken pains to present this material in an intellectually and morally responsible manner.[17]

(c) The process of *de-skilling* refers to a decline in the "tacit knowledge" required to put to use the sort of information referenced in (b). As the scholars Gautam Mukunda, Kenneth Oye, and Scott Mohr write, "Tacit knowledge consists of procedural and substantive knowledge primarily gained from experience instead of formal education," and it is "currently among the most significant barriers to bioweapons proliferation."[18] But as laboratory equipment becomes increasingly automated, less skill will be required to obtain the desired results. It is increasingly the case that one needs only a finger capable of pushing a button. This trend is arguably most notable with respect to synthetic biology, since this field is "*explicitly devoted* to the minimization of the importance of tacit knowledge," but it could apply even more strikingly to future nanotechnology, if it unfolds in the way that the nanotech pioneer Eric Drexler envisions (see below).[19]

(d) The *cost of the equipment* needed to build a makeshift laboratory is falling. This is being driven in part by the biohacker movement, which consists of amateur hobbyists who enjoy tinkering with the building blocks of life, and such falling costs could have the unintended consequence of making it far easier for terrorists to exploit advanced technologies for nefarious purposes. Furthermore, as we will discuss below, advanced nanotechnology could produce a huge range of technical artifacts for cheap, and the cost of weaponizing drones, for example, is falling at an exponential rate.

In sum, the tripartite cluster of dual usability, power, and accessibility is launching civilization into an era of distributed offensive capabilities that is genuinely unlike anything our species has ever before encountered. (As I have written elsewhere, "The malicious agents of the future will have bulldozers, rather than shovels, to dig mass graves for their enemies."[20]) Extrapolating the trends of (b) and (c) into the future, it is not implausible to imagine a world—perhaps years, decades, or centuries away—in which a large portion of the global population, or everyone, has access to WTDs.

4.2.2 Weapons of Total Destruction

We can distinguish here between two categories of *technogenic risks*, namely, **existing risks** and **emerging risks**. Roughly speaking, the former includes nuclear weapons and biotechnology, while the latter includes synthetic biology, nanotechnology, and tool AI.[21] As mentioned in section 1.4, the reason for considering the latter is that (a) the stakes are *so* astronomically high that we ought not dismiss too hastily any ideas proposed by reputable scholars in the relevant fields, and (b) if the emerging risks examined below become existing risks, they will introduce such behemoth hazards that thinking about them *now* would, to put it mildly, be prudent. So, let's look at these two categories in order.

(i) **Nuclear weapons**. These come in two varieties: atomic weapons, which use nuclear fission to produce an explosion, and hydrogen or thermonuclear weapons, which use both nuclear fission and fusion. Hydrogen bombs are much more destructive than their atomic siblings. For example, the first hydrogen bomb ever built, which the United States detonated in 1952, produced an explosion roughly 700 times greater than the atomic bomb dropped on Hiroshima.[22] Six years later, the Soviet Union detonated a hydrogen bomb called the "Tsar Bomba" that produced a blast around five times bigger than the United States' 1952 bomb. But this explosion could have been ten times worse "if not for the Soviets reducing its yield to limit the radioactive fallout."[23]

There are currently nine nuclear nations: the United States, the United Kingdom, Russia, France, China, India, Pakistan, North Korea, and Israel (although Israel refuses to acknowledge that it possesses any such weapons). The total number of nuclear weapons in the world as of this writing is approximately 15,000, and both the United States and Russia are in the process of modernizing their arsenals. Thousands of these weapons are still kept "on hair-trigger alert, ready to launch in under fifteen minutes," despite the Cold War having ended over two decades ago.[24] If the United States were to detect one or more missiles heading towards itself, "early warning crews manning their consoles 24/7 [would] have only three minutes to reach a preliminary conclusion" about what to do. If the crew members were to determine that a nuclear strike has been launched, "an emergency teleconference would be convened between the . . . President and his top nuclear advisers." The top officer would then "brief the president on his nuclear options and their consequences," an exchange that must take no longer than 30 seconds. The U.S. president would then have 12 minutes at most to decide whether nuclear retaliation is warranted.[25]

But inter-state conflicts aren't the only worry, of course: there is also the risk of nonstate actors using nuclear weapons against their perceived enemies.[26] For example, in a 2015 issue of their propaganda magazine, the Islamic State fantasizes about acquiring a nuclear weapon from Pakistan, which has a history of nuclear malfeasance—for instance, one of its leading nuclear scientists, A.Q. Khan, once sold secrets to Libya, Iran, and North Korea. After a weapon is obtained, the Islamic State writes,

> *The nuke and accompanying mujāhidīn arrive on the shorelines of South America and are transported through the porous borders of Central America before arriving in Mexico and up to the border with the United States. From there it's just a quick hop through a smuggling tunnel and hey presto, they're mingling with another 12 million "illegal" aliens in America with a nuclear bomb in the trunk.[27]*

The article adds that this scenario is "the sum of all fears for Western intelligence agencies and it's infinitely more possible today [in 2015] than it was just one year ago."[28]

As many scholars have noted, if even a *single* nuclear bomb were to explode anywhere around the world today, the consequences would be politically, economically, and psychologically devastating.[29] A RAND Corporation "scenario analysis," for instance, found that a nuclear weapon detonated in Los Angeles could not only kill 60,000 people instantly and expose some 150,000 more to radioactive contamination but spur millions to flee the region. Even more, "the economic effects of the catastrophe are likely to spread far beyond the initial attack, reaching a national and even international scale." Global trade could be severely disrupted, local labor supplies in port cities could dwindle, and the largest insurance companies in the country could go bankrupt. "While exact outcomes are difficult to predict," the report states, "these hypothetical consequences suggest alarming vulnerabilities. Restoring normalcy to economic relations would be daunting, as would meeting the sweeping demands to compensate all of the losses."[30]

According to the nuclear experts Gary Ackerman and William Potter, there was enough fissile material as of 2008 "for over 200,000 nuclear weapons." They add that "many of the sites holding this material lack adequate material protection, control, and accounting measures; some are outside the International Atomic Energy Agency's (IAEA) safeguard system; and many exist in countries without independent nuclear regulatory bodies or rules, regulations, and practices consistent with a meaningful safeguards culture."[31] In fact, the IAEA reports that

As of December 2013, a total of 2,477 incidents were reported to the IAEA's "Incident and Trafficking Database" (ITDB) by participating states. Of those, 424 incidents involved unauthorized possession and related criminal activities, or attempts to illegally trade or use nuclear material or radioactive sources. There were 664 reported incidents involving the theft or loss of nuclear or other radioactive material and a total of 1,337 cases involving other unauthorized activities, including the unauthorized disposal of radioactive materials or discovery of uncon-trolled sources.[32]

While a single detonation would have global repercussions—it would, indeed, change the course of history—the most extreme situation would involve an all-out exchange between two or more nuclear nations. When a nuclear weapon explodes, it can produce a shockwave strong enough to hemorrhage the lungs and abdominal cavity, initiate superhurricane-force winds, and spread radioactive fallout over a large area, which can cause acute radiation sickness and permanent damage to DNA. But multiple bombs exploding in high-density urban centers can result in *firestorms*, or massive conflagrations with sustained gale-force winds, that pollute the stratosphere with sunlight-blocking soot. This could induce a **nuclear winter** similar to the volcanic and impact winters previously discussed, leading to "significant surface darkening over many weeks, subfreezing land temperatures persisting for up to several months, large perturbations in global circulation patterns, and dramatic changes in local weather and precipitation rates."[33] Global agricultural systems would almost certainly collapse, followed by mass starvation, malnutrition, infectious disease, and other apocalyptic phenomena.

The president of the Ploughshares Fund, Joseph Cirincione, notes that since the "nuclear winter theory" was first proposed in the 1980s, "it has been repeatedly examined and reaffirmed."[34] For example, a 2007 study concludes that "even a small-scale, regional nuclear war could kill as many people as died in all of World War II and seriously disrupt the global climate for a decade or more, harming nearly everyone on Earth."[35] A regional exchange like this is not impossible, and perhaps not even that unlikely; for instance, India and Pakistan are both nuclear nations with a protracted history of conflict, and in 2002 tensions between them nearly culminated in a "war that both governments thought might go nuclear."[36] Neither India nor Pakistan are signatories of the Non-Proliferation Treaty, and the ongoing instability of Pakistan's government has worried many experts. Thus, as the Graham/Talent WMD Commission put it in a 2008 report, "were one to map terrorism and weapons of mass destruction today, all roads would intersect in Pakistan."[37]

Finally, in addition to atomic and hydrogen bombs, we should also note the possibility of neutron and cobalt bombs. The former was invented by the physicist Samuel Cohen to minimize the physical damage caused by nuclear weapons while maximizing the number of combatant casualties. It is a modified hydrogen bomb that produces a relatively small blast but releases large amounts of ionizing radiation in the form of free neutrons, which damages DNA. Similarly, the idea behind the cobalt bomb, once described by the physicist Leo Szilard as a "doomsday device," is to produce the largest radioactive fallout possible, thereby killing as many people as possible. This is achieved by adding cobalt metal to a hydrogen bomb. Upon detonation, the cobalt is converted into a radioactive isotope with a half-life of slightly more than five years—a long decay time that would make hiding in a shelter less practicable. To date, no cobalt bombs have probably ever been built, although this conclusion is based only on publicly accessible records.

(ii) **Biotechnology and synthetic biology.** These are overlapping domains of applied biology that aim to manipulate biological systems from the bottom-up. For example, genetic engineering uses biotechnology to modify organisms by either adding new genes or "knocking out" existing ones. Synthetic biology takes this a step further by building standardized "biological devices" that scientists can use to control cellular processes. At the extreme, this subfield strives to produce entirely synthetic organisms whose anatomical and physiological features are the direct product of human design.

The first synthetic organism on planet Earth, dubbed "Synthia," was created in 2010 by the geneticist Craig Venter and his team of scientists. They accomplished this incredible feat by modifying a genome on a computer (adding "markers" to it), synthesizing it from scratch in the laboratory, and then injecting it into a cell that subsequently replicated more than a billion times. Venter ultimately hopes to create a single-celled synthetic organism with the absolute minimal number of genes necessary for it to survive. Extra genes could then be added to make it behave in certain desirable ways—for example, to produce pharmaceuticals and biofuels or to extract carbon dioxide from the atmosphere (as a form of CDR).[38]

Another recent milestone of biotechnology/synthetic biology is the CRISPR/Cas9 system. This is a method of DNA editing that is based on the immune system of many prokaryotes—that is, bacteria and archaea.[39] It enables scientists to alter the DNA in cells—including both our somatic cells (those that make up most of our bodies) and germ cells (those that are passed down to future generations)—with extraordinary precision. This is why the *MIT Technology Review* named CRISPR/Cas9 one of the 10 breakthroughs of 2014 and 2016 and why *Science Magazine* gave it the "Breakthrough of the Year" award in 2015. As the geneticist Hugo Bellen enthusiastically exclaims, "Everything is possible with CRISPR. . . . I'm not kidding."[40]

The dual trends of power and accessibility are most pronounced, as of this writing, in the domains of biotechnology/synthetic biology, with both exciting and scary results. For example, a group of Australian scientists who were trying to create a mouse contraceptive by modifying the mousepox virus inadvertently made it 100 percent lethal in all mice, including those that had previously been vaccinated against it and those with a natural immunity. This confirms that making an already virulent virus *even more* virulent is possible with genetic engineering techniques, and it has bioterrorism implications because the smallpox and mousepox viruses are quite similar. A year later, in 2002, scientists at Stony Brook University synthesized "a live polio virus from chemicals and publicly available genetic information." Specifically, they created "the virus using its genome sequence, which is available on the Internet, as their blueprint and genetic material from one of the many companies that sell made-to-order DNA." This project was, in fact, funded by the Pentagon, and the point was "to send a warning that terrorists might be able to make biological weapons without obtaining a natural virus." As the study's lead scientist chillingly put it, "You no longer need the real thing in order to make the virus and propagate it."[41]

Other deadly pathogens that have been produced in the laboratory include "a SARS-like virus and the formerly extinct strain of the influenza virus" that caused the Spanish flu pandemic (see section 2.3).[42] Until recently, the only samples of the latter were preserved as "small DNA fragments in victims buried in Alaskan permafrost, or in tissue specimen of the United States Armed Forces Pathology Institute."[43] The purpose of resurrecting this virus in particular "was to gain insight into the genetic factors that made it so virulent, thereby guiding the development of antiviral drugs that would be effective against future pandemic strains of the disease." But doing so foregrounds the sinister possibility that malicious individuals could synthesize it as well. It also opens the door to an accidental release of a pathogen, which has happened before, with great consequences. For example, the 2009 swine flu outbreak was probably the result of a virus, preserved in laboratories since the 1950s, that was released by mistake in the late 1970s.[44] One study suggests that this outbreak may have killed upward of 203,000 people around the world.[45]

Finally, biotechnology and synthetic biology both present special difficulties for regulators. As the scientists Ali Nouri and Christopher Chyba write, "Biological weapons proliferation poses challenges more similar to those presented by cyber attacks or cyber terrorism than to those due to nuclear or chemical weapons. . . . Internet technology is so widely available that only a remarkably invasive inspection regime could possibly monitor it." Furthermore, whereas nuclear weapons require rare materials like enriched uranium and plutonium, biological microorganisms are self-replicating, some on timescales of "just twenty minutes, allowing microscopic amounts of organisms to be mass-produced in a brief period of time."[46] And certain pathogens, such as anthrax, can be found in one's backyard; a 2015 study even discovered traces of anthrax and the bubonic plague in New York City subways.[47]

What makes bioterrorism especially worrisome when compared to natural pandemics, which are themselves quite threatening, is that there is "some evidence for an inverse relationship between a pathogen's lethality and transmissibility," as a 2016 Global Challenges Foundation report observes.[48] That is to say, extreme virulence in nature will tend to lower the probability of successful propagation from one organism to another, since a germ that kills its host immediately will have fewer opportunities to spread. But as the 2016 report goes on to note, "Biotechnology has the potential to break this correlation, allowing organisms with extraordinarily high lethality and transmissibility."[49] Whereas a pathogen that combines, say, the lethality of rabies, the incurability of Ebola, the long incubation period of HIV, and the contagiousness of the common cold is unlikely to evolve through Darwinian natural selection, it could, at least in theory, be synthesized in the laboratory.[50] If this were to occur and the resultant pathogen was released in a busy airport, it could spread around the world without anyone noticing, due to delayed symptoms, and then suddenly cause a huge number of people to become deathly ill. Hospitals would be overwhelmed with patients, and doctors would be puzzled—not that a diagnosis would help, because there would be no cure. In a worst-case scenario, an **engineered pandemic** could cause civilization to collapse, perhaps never fully recovering to its pre-outbreak state.

Many risk scholars believe that a pandemic of this sort, enabled by biotechnology and synthetic biology, constitutes a profound near- and midterm threat to humanity, possibly equaling that posed by nuclear weapons.

(iii) **Molecular nanotechnology**. The term "nanotechnology" was coined by the Japanese scientist Norio Taniguchi in 1974 and subsequently popularized by Eric Drexler in his 1986 book *Engines of Creation*.[51] This neologism has proven over time to be rather elastic, though, and consequently what many contemporary scientists mean by "nanotechnology" isn't what Drexler has in mind. Whereas the former group tends to consider any nanoscale entity—such as nanoparticles in sunscreen or fuel—to count as nanotechnology, Drexler emphasizes manufacturing techniques capable of building high-quality products with *absolute atomic precision*.[52] This second, more revolutionary sense is now variously referred to as "atomically precise manufacturing," "molecular manufacturing," and "molecular nanotechnology," and it will be the focus of this section.

There are two types of anticipated molecular nanotechnology

that could pose major threats to humanity, namely, **nanofactories** and **self-replicating nanobots**. Taking these in turn: a *nanofactory* is a hypothetical device capable of manufacturing products with atomic precision by grabbing and repositioning individual atoms into specific locations. Thus, not only would two computers built by a nanofactory be identical with respect to their macroscopic properties—that is, their shape, color, hardness, weight, and so on—but if one were to "zoom in" to the submicroscopic level, one would find their corresponding particles in *exactly* the same places; i.e., they would be *structurally isomorphic down to the atom*.[53]

For a nanofactory to work, it would need three ingredients, namely, (1) some digital instructions, (2) a source of electrical power, and (3) a feedstock molecule, such as acetone or acetylene. The first will probably be free, downloadable from the Internet; the second affordable to nearly everyone, as it is today; and the third relatively cheap, given that this molecule could be bought in bulk.[54] Furthermore, a nanofactory need not be any larger than a few shoeboxes stacked together. Some theorists have, indeed, envisaged *personal nanofactories* (PNs) small enough to fit on top of one's desk in an office, or at home. This device could produce atomically precise components that users could assemble into a vast range of technical products, including smartphones, bicycles, household appliances, furniture, automobiles, ultrafast computers, and even orbital spacecraft. Perhaps the most incredible possibility here is that nanofactories could manufacture *other nanofactories*, thereby placing the means for almost unlimited material production into the hands of nearly everyone. As the nanotech experts Chris Phoenix and Mike Treder write, "The point at which nanofactories become able to build more nanofactories seems particularly noteworthy, because it is at this point that high-tech manufacturing systems could become, for the first time in history, non-scarce."[55]

The potential benefits of nanofactories are immense. Imagine a world in which a large percentage (or all) of humanity can satisfy their material needs by simply downloading some blueprints, plugging in a PN, and feeding it a source of simple molecules. Drexler describes this future in terms of "radical abundance," an idea echoed (and championed) by the futurists Ray Kurzweil and Peter Diamandis in their respective tomes.[56] But as with all dual-use technologies, the silver cloud has a dark lining: bad actors, both state and nonstate, could exploit nanofactories to produce massive arsenals of exceptionally dangerous weaponry, such as high-powered guns, bombs and missiles, self-guided bullets, aerospace materials that make aircraft undetectable, metamaterial cloaking, high-powered laser weapons, laboratory equipment for synthesizing designer pathogens, lethal insect-like drones, supercomputers on which to run dangerous AI systems, *and so on*.[57] Some scholars have even speculated that nanofactories could enable terrorist groups to produce nuclear weapons, although this is a topic of ongoing debate.

A society "armed to the teeth" with nanofactory-made weaponry could become dangerously unstable for multiple reasons. First, consider the implications for the stability of states, whose authority is predicated upon an asymmetrical power dynamic between them and their citizens.[58] Borrowing from the philosopher Thomas Hobbes, let us postulate that governments are the result of a *social contract* between individuals in a hypothetical "state of nature," which Hobbes imagines as a war of "all against all," where life is "solitary, poor, nasty, brutish, and short."[59] (Incidentally, contemporary anthropology partly corroborates this starting assumption.[60]) To escape the omnipresent threat of violence at the hands of one's neighbors, people enter into a social contract whereby the state provides security in exchange for individuals giving up some personal liberties. But here's the catch: this social contract only works if the state has the *capacity* to enforce laws, ordinances, agreements, contracts, and so on. If individuals become as powerful as the state, the social contract will dissolve, and with it the modern state system. This is one potential consequence of biotechnology, synthetic biology, and molecular nanotechnology: by distributing offensive capabilities across society, they could effectively "level the playing" field between states and citizens. The result would be a return to the Hobbesian plight of constant warring from which we came.[61]

Another issue concerns the potential for nanofactories to alter global trade relations. If nanofactories make each country self-sufficient, then we should expect global trade—or at least the trade of material goods—to decline. This is worrisome because research shows that "countries that [depend] more on trade in a given year [are] less likely to have a militarized dispute in the subsequent year, even controlling for democracy, power ratio, great power status, and economic growth."[62] In other words, the benefits of trade between two countries elevate the "conflict threshold" that must be exceeded for either to commit to going to war with the other. A related phenomenon is the Capitalist Peace theory, also known as the "Golden Arches" theory, which states that "no two countries with a McDonald's have ever fought in a war."[63] (The *only* clear exception to this rule is the 1999 NATO bombing of Yugoslavia during the Kosovo War.[64]) Thus, the dissolution of global trade between capitalist countries could increase the probability of martial confrontations. In Phoenix and Treder's words, "As economic interdependence disappears, a major motivation for partnership and trust also may be substantially reduced."[65] Depending on the conventional, chemical, biological, radiological, nuclear, and nanotech weapons that exist at the time, such conflicts could plunge civilization into an existential crisis.[66]

This leads to the second major threat from molecular nanotechnology. Whereas the atom-moving components of the nanofactory are stationary, nanobots are *autonomous mini-machines* capable of freely traversing their environments to produce some programmed effect. On the one hand, humanity could use such robots for highly beneficial ends, such as cleaning up the environment after a toxic spill. We could also design them to kill cancer cells, destroy the beta-amyloid plaques in the brains of Alzheimer's patients, and repair damaged organs after an injury. As the physicist Richard Feynman speculates in a 1959 talk that introduced the concept of nanotechnology,

Although it is a very wild idea, it would be interesting in surgery if you could swallow the surgeon. You put the mechanical surgeon inside the blood vessel and it goes into the heart and "looks" around. . . . It finds out which valve is the faulty one and takes a little knife and slices it out. Other small machines might be permanently incorporated in the body to assist some inadequately-functioning organ.[67]

On the other hand, though, agents with wicked intentions could exploit autonomous nanobots for existentially harmful ends. For example, consider the possibility of *self-replicating nanobots*, or nanobots purposively designed to convert all the organic matter that they come into contact with into copies of themselves. Imagine that someone drops a single nanobot of this sort into a deciduous forest. What would happen? Upon landing on a bed of leaves, moss, grass, and microbes, it would make a clone of itself from the organic stuff surrounding it. These two nanobots would then make copies of themselves, yielding four nanobots, and so on, resulting in an exponential explosion of the nanobot population until the entire continent (or group of contiguous continents) is covered by a wriggling swarm of mindlessly reproducing mini-machines. If such nanobots were blown over the ocean by the wind (like pollen) or intentionally transferred to another region, the biosphere *in toto* could be destroyed—an "ecophagic" disaster referred to as the **grey goo scenario**.[68]

Or consider a doomsday scenario outlined by Ray Kurzweil:

In a two-phased attack, the nanobots take several weeks to spread throughout the biomass but use up an insignificant portion of the carbon atoms, say one out of every thousand trillion (10^{15}). At this extremely low level of concentration the nanobots would be as stealthy as possible. Then, at an "optimal" point, the second phase would begin with the seed nanobots expanding rapidly in place to destroy the biomass. For each seed nanobot to multiply itself a thousand trillionfold would require only about fifty binary replications, or about ninety minutes. With the nanobots hav-

ing already spread out in position throughout the biomass, movement of the destructive wave front would no longer be a limiting factor.[69]

Even more insidiously, one could, in theory, design autonomous nanobots to target specific human races or biological species. This could be accomplished by programming them to attack an organism if and only if they recognize particular genetic signatures that are unique to the relevant group. If spread across a sufficiently wide area, the result could be a genocidal or, worse, omnicidal catastrophe.[70]

While nanofactories and nanobots remain speculative, many reputable scholars expect them to become a reality later this century.[71] Given the unprecedented risks that this category would introduce, it is not too soon, one could argue, to start thinking seriously about how best to avoid a worst-case outcome.

(iv) **Tool AI**. This category poses a more debatable existential risk than those discussed above. Nonetheless, there are some reasons for including it in this book. Consider the following scenario proposed by the computer scientist Stuart Russell:

> *A very, very small quadcopter, one inch in diameter can carry a one- or two-gram shaped charge. You can order them from a drone manufacturer in China. You can program the code to say: "Here are thousands of photographs of the kinds of things I want to target." A one-gram shaped charge can punch a hole in nine millimeters of steel, so presumably you can also punch a hole in someone's head. You can fit about three million of those in a semi-tractor-trailer. You can drive up I-95 with three trucks and have 10 million weapons attacking New York City. They don't have to be very effective, only 5 or 10% of them have to find the target.*[72]

This scenario could be scaled up arbitrarily: perhaps a rogue state packs *100 million* of these weapons into *hundreds* of semitrucks around the world and then deploys this drone army within a five-minute window. The resulting devastation could have similar effects to a nuclear war or global pandemic by disrupting the global economy, causing mass panic, and initiating retaliatory wars. Such an attack could also be perpetrated by nonstate actors, given the power and accessibility trends previously outlined. Thus, Russell adds that "there will be manufacturers producing millions of these weapons that people will be able to buy just like you can buy guns now, except millions of guns don't matter unless you have a million soldiers. You need only three guys to write the program and launch [the drones]."[73]

Before ending this section, it is worth looking closer at the term "tool AI." The crucial idea here is that agency is a spectral rather than binary property. On one end of the spectrum are objects like rocks, which have no agency whatsoever. In the middle of the spectrum one finds artifacts like heat-seeking missiles (mentioned in section 1.3), which exhibit agency to the degree that they can navigate the physical world on their own to hit their targets. But heat-seeking missiles don't choose their targets. As the techno-philosopher Peter Asaro writes about the U.S.'s military drones, "These combat aircraft are capable of numerous sophisticated automated flight processes, including fully automated take-off and landing, GPS waypoint finding, and maintaining an orbit around a GPS location at a designated altitude, as well as numerous automated image collection and processing capabilities." Yet they "are not considered to be autonomous because they are still operated under human supervision and direct control."[74] It is, indeed, humans who make the morally important decisions about who to kill, when this should happen, and how it should take place. Thus, on the other end of this spectrum are *fully autonomous* systems capable of making a wide range of independent decisions to achieve their goals, whatever those are. An artificial general intelligence (AGI) would be an instance of this, which is why machine superintelligence is more appropriately classified as an "agential risk" (see subsection 4.3.1). In fact, one reason that superintelligence constitutes such an immense threat to humanity is precisely because of its autonomy—a property that could place it entirely outside human control.

In sum, as AI technology advances, the use of drones or other robots with semi-autonomous capabilities could pose an increasingly significant threat. As Ariel Conn of the Future of Life Institute writes, "We need to figure out how to deal with people who will do bad things with good AI systems and not just worry about AI that goes bad."[75]

4.3 Agential Risks

> *All else being equal, not many people*
> *would prefer to destroy the world.*
> *—Eliezer Yudkowsky*[76]

Let's begin by defining an "agential risk" as follows:

An agential risk refers to any agent who could pose a threat to humanity or human civilization if she or he were to gain access to a WTD.

There are two types of agential risk, depending on the agent's *proximate motivational state*: **agential terror** and **agential error**. The former denotes scenarios in which the relevant agent *intends* to cause existential harm, whereas the latter involves an agent who causes existential harm *by accident*. We shall examine these in order.

4.3.1 Agential Terror

Imagine that a **doomsday button** were suddenly placed in front of every person alive on Earth right now. If pushed, it would initiate a world-destroying WTD, resulting in one of the four existential risk scenarios outlined by Bostrom. Call this the *doomsday button test*. Thus, the empirically interesting question is: Who exactly would push this button? Who would "pass" the doomsday button test by deliberately destroying the world?

Consider first the Provisional Irish Republican Army (PIRA), which has been responsible for numerous terrorist attacks against the United Kingdom in an attempt to take back Northern Ireland. If WTDs were available to PIRA during the height of conflict, between roughly 1971 and 1994, would this group have employed them? Almost certainly *not*, because its political goals were predicated upon the *continued existence* of human civilization. PIRA wanted to change the world, not destroy it. Now, consider al-Qaeda, the Islamist group responsible for the 9/11 terrorist attacks that killed nearly 3,000 Americans.[77] Although its then-leader, Osama bin Laden, once called it his "religious duty" to acquire WMDs, the group was motivated by religio-political grievances, such as the U.S. military presence in Saudi Arabia and Western sanctions on Iraq, which resulted in an estimated 500,000 excess childhood deaths, according to a 1999 UNICEF report.[78] So, as with PIRA, bin Laden's goal was to reconfigure the world (by defeating the West), not to bring about its end. Finally, consider a ruthless autocrat who dreams of ruling a global totalitarian state. How tempted would such an individual be to press a doomsday button in service of this goal, if one were available? Given that an autocrat can't rule the world if the world doesn't exist, once again it appears unlikely that she or he would cause a techno-apocalypse on purpose (although see Box 7).

There are some agents, though, who would reliably pass the doomsday button test.[79] These are:

Box 7. With respect to autocrats, the philosopher Nicholas Agar points out that "dictators are susceptible to tantrums when they feel that their just and noble aims are thwarted."* Consequently, some dictators may list the following outcomes of conflict from best to worst: (i) victory, (ii) the total destruction of civilization, (iii) losing. We might therefore anticipate that lost ground during a conflict could push an autocrat with access to WTDs to consider using them for world-destroying purposes. A similar line of reasoning could apply equally well to politically motivated terrorist groups: some might prefer annihilation over defeat even while wishing for neither.

* Personal communication

(i) **Apocalyptic terrorists.** Let's begin with some background information. According to the 2016 Global Terrorism Index, religious extremism is the primary driver of global terrorism today.[80] Although many of the earliest forms of terrorism were religious— such groups gave us words like "zealot," "thug," and "assassin"— the nineteenth and twentieth centuries were dominated by terrorists motivated by nationalist, separatist, anarchist, Marxist, and other secular ideologies. This began to change in the late 1980s with the rise of Islamic terrorism, a trend that culminated with the 9/11 atrocities and the subsequent formation of the Islamic State in Iraq and Syria, Boko Haram in Nigeria, and numerous Shia militias throughout the Middle East. What is notable about religious terrorism, in particular, is that it is much more lethal and indiscriminate than past forms of political terrorism, often intentionally choosing "soft targets" to maximize civilian casualties and media exposure.[81] (In a sense, terrorism is a form of communication that has been given a global megaphone by modern media.) Today, the two most dangerous terrorist groups are Boko Haram and the Islamic State—the latter of which we will derogatorily refer to as "Daesh."[82]

Apocalyptic terrorism is the most radical form of religious terrorism and therefore the most dangerous. The apocalyptic terrorist believes that (a) the apocalypse is imminent, and (b) she or he has a special role to play in bringing about this event, often using violence as the catalyst. To borrow a line from the former CIA director Jim Woolsey, such terrorists "are not seeking a place at the table, but are seeking to blow up the table and kill everyone sitting there." Indeed, they see the struggle as metaphysically transcendental in nature. It constitutes the *ultimate battle* between Good and Evil—a one-time epic clash of cosmic opposites that has only a single possible outcome, namely, the complete obliteration of God's enemies. When deeply held, as they sometimes are, these convictions can produce a grandiose sense of moral righteousness and eschatological urgency that true believers can use to justify—in their own minds—nearly any act of brutality and carnage, no matter how catastrophic.[83] In a phrase, apocalyptic terrorists believe that *the world must be destroyed to be saved*, and that it is their divinely mandated task to ensure the former.[84]

Although rarely acknowledged, even in academic monographs, the bloody roads of history are littered with movements that were animated by "active apocalyptic" ideologies (see section 1.7). For example, the Taiping Rebellion was a 14-year-long conflict between the Taiping Heavenly Kingdom and the Qing dynasty in China led by a charismatic apocalypticist named Hong Xiuquan. Hong believed that he was the younger brother of Jesus, and his teachings combined elements of Buddhist, Confucian, and Christian eschatologies. The result was a syncretistic worldview that fueled a monumental struggle—indeed, the *second most deadly* conflict in human history, resulting in some 30–35 million deaths.[85]

On the other side of the planet, between the 1930s and the 1980s, an apocalyptic ideology known as "Christian Identity" morphed into an influential doctrine among Christian racists and other fringe believers. According to Christian Identity, white Europeans are the true Israelis and the Jewish people are impostors who were literally born of Satan.[86] Furthermore, the end of the world is nigh, but it will not commence until white Christians "wage a great battle on the side of God against Satan, the Jews, and people of color," ultimately purifying the world of other races, the "mud people," through the use of catastrophic violence.[87] The Christian Identity movement has shaped the ideologies of right-wing groups like the Ku Klux Klan, the Aryan Nations, and The Covenant, the Sword, and the Arm of the Lord (CSA). In fact, the CSA planned an attack (that never occurred) very similar to the 1995 Oklahoma City bombing while they were literally "training 1,200 recruits in the Endtime Overcomer Survival Training School."[88] Given the rise of far-right movements in the Western world, as of this writing, the Christian Identity movement or some variation could reemerge as a formidable threat in the future.

Another example comes from the Japanese doomsday cult Aum Shinrikyo. This group perpetrated the worst terrorist attack in Japanese history when, in 1995, they released sarin gas into the Tokyo subway system, killing 12 people and injuring almost 5,000 others. What is particularly notable about this attack is that it was explicitly intended to trigger the battle of Armageddon—or World War III—which members of Aum believed would cause the total destruction of humanity, except for the group. Aum was also responsible for a failed 1993 bioterrorism attack involving anthrax (they sprayed mists of anthrax from a building and a van, but no one was sickened), and when their compound was raided police found enough chemicals to produce quantities of sarin that could kill 4 million people. Interestingly, many of Aum's followers were highly educated in fields of science and engineering.

Lastly, consider Daesh (the Islamic State). Although not all of Daesh's fighters are ardent apocalypticists, its leadership almost certainly consists of "true believers," to quote the journalist Graeme Wood, who genuinely embrace an imminent eschatology.[89] The first leader of what later became Daesh—a sadistic psychopath named Abu Musab al-Zarqawi—was fond of mentioning the battle of Armageddon between the Muslim forces and the "Romans," which prophetic hadith say will occur in the small Syrian town of Dabiq (section 1.7).[90] Thus, al-Zarqawi declared before his death in 2006 that "the spark has been lit here in Iraq, and its heat will continue to intensify—by Allah's permission—until it burns the crusader armies in Dabiq." The next leader of Daesh, Abu Ayyub al-Masri, was convinced that the end-of-days messianic figure, the Mahdi, would soon appear in Iraq, and consequently he made a number of strategic decisions (that backfired) based "on an apocalyptic timetable." As Will McCants documents, when al-Masri was criticized for these decisions, he simply replied, "The Mahdi will come any day."[91]

The current caliph of Daesh is Abu Bakr al-Baghdadi, a scholar who earned a PhD in Islamic Studies from a reputable university (i.e., Baghdad University). Since the Mahdi didn't show up when al-Masri expected, Daesh's apocalyptic focus shifted under al-Baghdadi to the establishment of a caliphate, which a hadith prophesies will reform before the Last Hour.[92] Still, febrile anticipation of the Grand Battle in Dabiq continues, and Daesh fighters have frequently tried to coax the United States into sending troops to Syria so that Armageddon can begin.[93] This is why a number of beheading videos were filmed in or around Dabiq, with one executioner ("Jihadi John") saying, "Here we are, burying the first American Crusader in Dabiq, eagerly waiting for the remainder of your armies to arrive."[94]

If any of these groups were to gain access to a doomsday button, there is a high probability that they would push it. The outcome could be permanent stagnation or human extinction, although one might argue that the latter is, all things considered, more probable. The reason, as Box 10 explores, is that religious eschatologies reject the very possibility of human extinction; rather, they specify some community of believers who will *survive* the cataclysmic paroxysms of the eschaton.[95] (The issue is further complicated by the ontological commitment of dualistic religions to *immortal souls* that can exist without the physical body.) This conviction—that human extinction is impossible—could lead apocalyptic terrorists to be overly *careless* with extinction-causing WTDs. For example, they might induce a global catastrophe thinking that it will usher in the religious utopia described by their holy books—that is, thinking that some portion of humanity will be saved—when in reality this event will almost certainly trip our species into the eternal grave. (Imagine a highly lethal germ being released around the world on the assumption that followers of such and such a creed will be protected by God.) The combination of *the will to cause mass destruction* and *a belief that mass destruction could only be so bad (and ultimately good)* could lead such individuals to severely underestimate the consequences of their actions.[96]

(ii) **Misguided ethicists**. What concerns us here are the potential consequences of certain moral theories on the specific goal of attaining "desirable future development."[97] Let's set the stage with a view called "antinatalism." This was espoused by the German philosopher Arthur Schopenhauer, and its most vigorous contemporary defender is the South African philosopher David Benatar. The central idea is that "coming into existence is always a serious harm." As Benatar writes,

> *Although the good things in one's life make it go better than it otherwise would have gone, one could not have been deprived by their absence if one had not existed. Those who never exist cannot be deprived. However, by coming into existence one does suffer quite serious harms that could not have befallen one had one not come into existence.[98]*

The conclusion is that it is *morally wrong* to procreate, to bring new people into the world; it is better for everyone "never to have been." While this theory is not widely accepted by moral philosophers today, it does have some supporters. In addition, it is impossible to predict how the ethical landscape of the future might evolve, and if antinatalism were to become sufficiently widespread, it could threaten the perpetuation of our species.

But it does not engender an agential risk *per se*, on the definition outlined above. For this, we can turn to an ethical system known as "negative utilitarianism" (NU).[99] There are several versions of NU, some of which do not appear risky. For example, the Australian ethicist and adventurer Roger Chao argues for what he labels "negative average preference utilitarianism." According to this view, moral action should always aim to reduce the total amount of frustrated preferences in the world.[100] (It is thus closely related to *antifrustrationism*.) In contrast, "strong" negative utilitarianism (SNU for short) claims that moral action should always and entirely focus on *the elimination of suffering*.[101] Whereas classical utilitarianism emphasizes both pleasure and pain, happiness and sorrow, SNU emphasizes only pain and sorrow. As the philosopher David Pearce writes, negative utilitarianism arises

> *from a deep sense of compassion at the sheer scale and intensity of suffering in the world. No amount of happiness or fun enjoyed by some organisms can notionally justify the indescribable horrors of Auschwitz. Nor can it outweigh*

*the sporadic frightfulness of pain and despair that occurs
every second of every day.*[102]

It follows that, from the SNU perspective, a world full of
overwhelming bliss *plus a single pinprick* is morally worse than
a world in which there exists neither bliss nor a pinprick; the
only relevant factor for assessing which world is better is the total
amount of suffering therein. It is this line of reasoning that led
the philosopher R.N. Smart to propose, in a 1958 article, the most
famous criticism of negative utilitarianism: it seems to entail that
one should endorse a "world-exploder" who annihilates all sentient life in the universe, since doing so would eliminate every instance of suffering. Yet, as he puts it, "we should assuredly regard
such an action as wicked."[103]

Although negative utilitarianism has yet to gain a large following, there are perhaps "a few hundred—or at most a few
thousand—persons scattered across the globe [who] currently
acknowledge the NU title."[104] But it is not inconceivable that SNU
gains popularity in the coming years or decades, especially if one
or more major catastrophes cause intense human suffering that
media outlets thrust into everyone's perceptual field through
TVs, smartphones, and other media (see below). Just think of
how the Vietnam War coverage fueled anti-war sentiment in the
United States, or the image of Alan Kurdi, a three-year-old Syrian boy who drowned while trying to cross the Mediterranean
Sea, has inspired pro-immigration activism in recent years.[105] To
reverse the common aphorism, "Within sight, within mind," an
ostensible fact that could grow the ranks of SNU.

But Pearce, himself a negative utilitarian, responds to criticisms of his view by arguing that *classical utilitarianism* itself
could pose an existential threat to humanity.[106] In his words, "A
thoroughgoing classical utilitarian is obliged to convert your matter and energy into pure utilitronium, erasing you, your memories, and indeed human civilisation."[107] The term "utilitronium"
signifies a configuration of matter and energy that optimizes total
utility; it is an organized state of physical stuff capable of realiz-

ing far more well-being than the human organism. Thus, Pearce claims that classical utilitarians are

> obliged to erase such a rich posthuman civilisation with a utilitronium shockwave. . . . The "shockwave" in utilitronium shockwave alludes to our hypothetical obligation to launch von Neumann probes propagating this hyper-valuable state of matter and energy at, or nearly at, the velocity of light across our Galaxy, then our Local Cluster, and then our Local Supercluster.[108]

Strong negative utilitarians and utilitronium shockwave advocates (insofar as the latter exist, or should exist, as Pearce suggests) constitute interesting subtypes of agential risks because *outside* of the context in which world-destroying WTDs are available, they pose no threat to humanity. Indeed, quite the opposite: negative utilitarians wish to reduce suffering while classical utilitarians prefer a world marked by minimal suffering and maximal happiness. *Only* once WTDs become available to the former and *only* once von Neumann probes (capable of converting exoplanets into utilitronium) become available to the latter would these agents pose an extinction risk. Since both WTDs and von Neumann probes could become available in the future, risk scholars should monitor, in whatever way appropriate, "misguided ethicists" in the future.

(iii) **Idiosyncratic actors**. This category includes agents with idiosyncratic motives to destroy civilization or humanity—that is, motives sufficiently unique to the agent as not to fall within any other category here listed. Rampage killers and school shooters provide a paradigm case of the relevant mindset: in some instances, they have simply wanted *to kill as many people as possible and then die*, thus "going out with a bang." This was the situation with Eric Harris and Dylan Klebold, who perpetrated the 1999 Columbine High School massacre. As Harris, the mastermind behind this attack, scribbled in his personal journal, "If you recall your history the Nazis came up with a 'final solution' to the Jewish problem. Kill them all. Well, in case you haven't figured it out yet, I say 'KILL MANKIND' no one should survive."[109] He also wrote that "I think I would want us to go extinct," adding, "I just wish I could actually DO this instead of just DREAM about it all," and "I have a goal to destroy as much as possible . . . I want to burn the world."[110] Elsewhere he declared,

> *If I can wipe a few cities off the map, and even the fuckhead holding the map, then great. Hmm, just thinking if I want all humans dead or maybe just the quote-unquote "civilized, developed, and known-of" places on Earth, maybe leave little tribes of natives in the rain forest or something. Hmm, I'll think about that.*[111]

After nearly killing 488 students with an improvised bomb made out of a propane tank, Harris—wearing a shirt with the phrase "Natural Selection"—and Klebold murdered 12 of their peers and one teacher and then committed suicide. The Columbine massacre was the deadliest school shooting in American history until Adam Lanza killed 20 children and 6 adults at Sandy Hook Elementary School in 2012 before taking his own life. It was also inspired by the notorious mass murderer Charles Manson, who, incidentally, believed in an impending apocalyptic race war (which he dubbed "Helter Skelter") and once wrote that "I'm going to kill as many of you as I can. I'm going to pile you up to the sky. I figure about fifty million of you."[112] If a future misanthrope with grandiose *murder-suicide* inclinations were to gain access to a doomsday button hooked up to nuclear weapons, designer pathogens, ecophagic nanobots, or swarms of lethal drones, it is very likely that she or he would push it. This would, indeed, enable one to "go out with the *ultimate bang*," thereby fulfilling Harris's dark fantasy of "leav[ing] a lasting impression on the world."[113]

Scholars have identified a number of environmental and psychiatric variables associated with school shooter types. For example, a lack of parental supervision and family problems like divorce, as well as social "isolation and rejection from peers and teachers," are contributing factors.[114] Other issues include access to firearms (in our case, WTDs), situations that impede feelings of self-esteem, a recent personal loss or episode of humiliation, and bullying—although some school shooters are best described as "bully-victims" who both bullied others and were themselves bullied.[115] They also tend to suffer from mental or personality disorders like depression, schizophrenia, schizotypal personality disorder, narcissistic personality disorder, and/or sociopathy, the last of which is associated with impaired empathy, egotistical and egocentric traits, and antisocial behavior.[116] Incidentally, the psychologist Martha Stout estimates that approximately 4 percent of the population consists of sociopaths who, by virtue of their condition, lack a *conscience*, or "the inner sense of what is right or wrong in one's conduct or motives."[117] As Stout puts it, imagine having

> *no feelings of guilt or remorse no matter what you do, no limiting sense of concern for the well-being of strangers, friends, or even family members. Imagine no struggles with shame, not a single one in your whole life, no matter what kind of selfish, lazy, harmful, or immoral action you had taken.[118]*

If Stout's estimate is accurate, then there are about 300 million sociopaths in the world today, and we should expect roughly 372 million by 2050, if the global population rises to 9.3 billion. Although not all sociopaths are violent, they make up a disproportionate segment of the prison population, about 20 percent in the United States. Even more, they "account for more than 50 percent of the 'most serious crimes' (extortion, armed robbery, kidnapping, murder) and crimes against the state (treason, espionage, terrorism)."[119] There are also reasons for suspecting that many dictators throughout history have been sociopaths, an issue that ties into the phenomenon of bad governance, explored separately in chapter 5.

Another exemplar of this category is Marvin Heemeyer, a Colorado welder who owned a muffler repair shop. Heemeyer became embroiled in disputes with his town over zoning issues and had to pay several thousand dollars in fines for property violations. In retaliation for what he saw as unfair treatment, he converted a large bulldozer into a "futuristic tank" complete with armor, mounted video cameras, and three gunports.[120] On June 4, 2004, Heemeyer got into the tank and started moving toward town. With a top speed of a slow jog and numerous police walking behind him during the incident, he proceeded to destroy one building and vehicle after another. Neither a flash-bang grenade thrown into the bulldozer's exhaust pipe nor 200 rounds of ammunition stopped him. After more than two hours of relentless destruction, the bulldozer became lodged in a basement, at which point Heemeyer shot himself with a pistol. Call this incident "Heemeyer's rampage."[121]

What is interesting about this, when juxtaposed with school shooter cases, is that Heemeyer didn't injure anyone, and this *might* have been intentional.[122] His primary mission, it appears, was merely to cause as much physical damage to the town as possible. If this is true, he provides a template for someone who might opt to use a WTD not to cause human extinction *per se*—as would likely have been the case with Harris and Klebold—but to destroy civilization, thus realizing a permanent stagnation or, if we have already reached a posthuman state, subsequent ruination scenario.[123]

Other notable incidents within this category include the Luby's massacre, the San Ysidro McDonald's massacre, the Bath School disaster, the University of Texas tower shooting, and the Tsuyama massacre.[124]

(iv) **Future ecoterrorists.** According to the "deep ecology movement," founded by the Norwegian philosopher Arne Næss in the 1970s, the natural world possesses intrinsic value independent of its instrumental value to humans and therefore it is worth saving "for its own sake." The *biospheric egalitarianism* of this perspective contrasts with the *anthropocentrism* of what Næss calls the "shallow ecology movement," which exhorts humanity to "fight against pollution and resource depletion" for the sole purpose of ensuring "the health and affluence of people in the developed countries."[125]

If one accepts biospheric egalitarianism and acknowledges that human activity is irreversibly destroying the environment, it takes only one additional (dubious) step to reach the conclusion that *Homo sapiens*, the self-described "wise man," must be exterminated. By analogy, imagine that *Periplaneta americana,* the American cockroach, were almost entirely responsible for climate change and the sixth mass extinction. If this were the case, humanity would most assuredly launch an all-out war on the cockroach that would conclude only once the entire species had been wiped out. But if humans are no more intrinsically valuable than the cockroach, the exact same logic should apply to us.[126] One should therefore advocate human extinction.

There are, in fact, several movements that advocate precisely this. For example, the Voluntary Human Extinction Movement (VHEMT, pronounced "vehement") claims to present "an encouraging alternative to the callous exploitation and wholesale

Source: Created by Nina Paley with Les U. Knight, Voluntary Human Extinction Movement, www.vhemt.org

destruction of Earth's ecology." Its website states that "when every human chooses to stop breeding, Earth's biosphere will be allowed to return to its former glory, and all remaining creatures will be free to live, die, evolve (if they believe in evolution), and will perhaps pass away, as so many of Nature's 'experiments' have done throughout the eons."[127] But as VHEMT emphasizes, often with an entertaining dose of levity, this goal must be accomplished *without coercion or violence*. Thus, it does not advocate terroristic tactics to bring about human destruction, and as such its supporters (just like antinatalists) are not agential risks on our definition.[128]

But this is not the case with other groups and individuals. Consider the Finnish deep ecologist and fisherman Pentti Linkola, who was one of Finland's "most celebrated" authors in the 1990s.[129] Linkola believes that Western society is guilty of a perverse "overemphasis on the value of human life," that "on a global scale, the main problem is not the inflation of human life, but its ever-increasing, mindless over-valuation."[130] He claims that another world war would be "a happy occasion for the planet," and suggests that, to avoid an ecological catastrophe, "some transnational body [or] small group equipped with sophisticated technology and bearing responsibility for the whole world" should attack "the great inhabited centres of the globe."[131] He has also avowed, rather eerily, that "if there were a button I could press, I would sacrifice myself without hesitating, if it meant millions of people would die."[132] And while he acknowledges that the environmental situation is worsening by the day, Linkola reassures his followers (in 1994) that "we *still* have a chance to be cruel. But if we are not cruel today, all is lost."[133]

Similarly, the lone-wolf ecoterrorist James Lee, who once held three people hostage at the Discovery Channel building "with explosives strapped to his body and a gun in his hand," argues that "children represent FUTURE catastrophic pollution whereas their parents are current pollution. NO MORE BABIES!," adding, "the humans? The planet does not need humans."[134] And the Gaia Liberation Front (GLF) writes in its "Statement of Purpose

(A Modest Proposal)" that humanity is an "alien species," "virus," or "cancer" that must be excised from the planet. It claims that doing this through nuclear war would result in too much collateral damage, mass sterilization is too slow, and suicide is impractical; but bioengineering offers "the *specific* technology for doing the job right—and it's something that could be done by just one person with the necessary expertise and access to the necessary equipment." As they write,

> Genetically engineered viruses . . . have the advantage of attacking only the target species. To complicate the search for a cure or a vaccine, and as insurance against the possibility that some Humans might be immune to a particular virus, several different viruses could be released (with provision being made for the release of a second round after the generals and the politicians had come out of their shelters).[135]

Along these lines, a 1989 article in the *Earth First! Journal* importunes the following:

> Contributions are urgently solicited for scientific research on a species specific virus that will eliminate Homo shiticus from the planet. Only an absolutely species specific virus should be set loose. Otherwise it will be just another technological fix. Remember, Equal Rights for All Other Species.[136]

Both these excerpts specifically single out weaponized biology: designer bugs that could wipe out the human species. But as the emerging technologies of subsection 4.2.2 reach fruition, ecoterrorists may find *non-biotech* weapons even more attractive. For example, self-replicating nanobots that target *Homo sapiens* wouldn't be subject to genetic mutations and thus could potentially offer a more reliable way of satisfying the "species-specific" condition. A vial of nanobots released in a few major urban centers could initiate a nearly unstoppable human extinction event that, at least in theory, would minimally disrupt natural ecosystems. Ecoterrorists could also try to discharge a deadly horde of AI drones in multiple megacities around the globe in an attempt to cripple modern society. This is an issue that we will return to at the end of subsection 6.3.3.

Similar to the school shooter/Heemeyer distinction made above, there are also cases of radical eco-anarchists, anarcho-primitivists, and neo-Luddites who would preferentially use WTDs to catapult humanity "back to the Pleistocene," as the primitivist slogan goes, rather than (intentionally) causing our extinction.[137] Consider Ted Kaczynski, the Unabomber. His primary complaint was that industrial civilization has severely compromised human freedom; thus, only once humanity embraces small, sustainable communities with rudimentary artifacts will freedom once again flourish. This thesis borrows much less from the deep ecology movement than from a strain of Ludditic thought associated with technology critics like Jacques Ellul and Lewis Mumford, the latter of whom once wrote that "if we are to prevent megatechnics from further controlling and deforming every aspect of human culture, we shall be able to do so only with the aid of a radically different model derived directly, not from machines, but from living organisms and organic complexes (ecosystems)."[138] To draw attention to his cause, Kaczynski began a campaign of domestic terrorism in 1978, during which he sent bombs to airlines and universities (indeed, Noam Chomsky was on his hit list), ultimately killing three and injuring twenty-three others. He later used this as leverage to get the *New York Times* and *Washington Post* to publish a 35,000-word manifesto called "Industrial Society and Its Future." Shortly afterward, Kaczynski's brother recognized his writing and alerted the FBI, thereby leading to his arrest.[139]

The point is that Kaczynski was driven not by a *death wish for humanity* but by a *destruction wish for civilization*. His goal was to transition global society to the "positive ideal" of, in his words, "WILD nature."[140] As he wrote in the 1995 manifesto,

We therefore advocate a revolution against the industrial system. This revolution may or may not make use of violence; it may be sudden or it may be a relatively gradual process spanning a few decades. . . . Its object will be to overthrow not governments but the economic and technological basis of the present society.[141]

In the years since his capture, Kaczynski has invigorated other violent groups to target people seen as guilty participants in the baneful techno-industrial system. For example, a Mexican group called "Individualidades Tendiendo a lo Salvaje" (ITS), meaning "Individuals Tending to the Wild (or Savagery)," sent a mail bomb to the Monterrey Institute of Technology and Higher Education in Mexico City in 2011 that seriously injured a robotics researcher and ruptured the eardrum of a computer scientist. As the *Chronicle of Higher Education* reports, ITS has also "been linked to attacks in France, Spain, and Chile," and the group—which has a nominal presence across much of Latin America—took responsibility for the murder of an engineering student, Lesvy Rivera, at the National Autonomous University of Mexico on May 3, 2017.[142] While ITS was initially "real slavish" to Kaczynski, it appears to have more recently adopted an omnicidal ecofascist ideology according to which "the human being deserves extinction."[143]

Although there have been few notable cases of environmental terrorism in the past decade and a half, we will see in subsection 4.3.3 that this may change in the coming decades. For the remainder of this book, I will refer to the cluster of overlapping but non-identical groups relevant to this category—including deep ecology extremists, radical environmentalists, eco-fascists, anticivilization fanatics, violent technophobes, anarcho-primitivists, militant neo-Luddites, and fringe eco-anarchists (or green anarchists)—using the single, imprecise appellation "ecoterrorists."[144]

(v) **Machine superintelligence**. Due to a *hardware bias* according to which agents composed of artificial rather than biological materials are given special attention among researchers, machine superintelligence is the only agential risk that has been studied in detail by those who one could locate under the umbrella of "existential risk studies." But this is beginning to change, and the present book hopes to be a catalyst for a more inclusive analysis of risky agents. Nonetheless, many experts concur that superintelligence poses an exceptionally grave danger to our collective future. Some, including myself, believe that it constitutes the greatest known threat to our long-term survival in the universe.

The reasons for this opinion are complex. Many people unfamiliar with the topic envision a cinematic battle between humans and a belligerent army of robotic foes, often bipedal androids with glowing red eyes and machine guns who for some reason wish to exterminate humanity. But this is *not* the kind of scenario that AI risk scholars worry about—indeed, evil androids are among the top nine "myths about advanced AI" (see Figure E).[145] Rather, the dangers are *far more menacing*, associated with what scholars call the **control problem**. This section considers some of the central issues that make this type of hardware-based agential risk so very risky.

(1) *The orthogonality thesis.* One can distinguish between two types of rationality, namely, *instrumental rationality* and *value rationality*. An agent is "instrumentally rational insofar as she adopts suitable means to her ends," whereas an agent is "value-rational" insofar as her ends are consciously and deliberately chosen for moral, epistemic, aesthetic, or other reasons.[146] Thus, an agent can be instrumentally rational without being value-rational, and vice versa. Within the fields of philosophy and cognitive science, the concept of *intelligence* is roughly synonymous with instrumental rationality and quite unrelated to value rationality. An agent is thus intelligent to the extent that it can acquire effective means to achieve its goals, whatever they happen to be.

With this in mind, the orthogonality thesis states that "intelligence and final goals are orthogonal axes along which possible agents can freely vary. In other words, more or less any level of intelligence could in principle be combined with more or less any

Figure E. AI: Myths vs. Facts

Myth/Mythical worry	Fact/Actual worry
Myth: Superintelligence by 2100 is inevitable **Myth:** Superintelligence by 2100 is impossible	**Fact:** It may happen in decades, centuries or never: AI experts disagree & we simply don't know

Myth: Only Luddites worry about AI	**Fact:** Many top AI researchers are concerned

Mythical worry: AI turning evil **Mythical worry:** AI turning conscious	**Actual worry:** AI turning competent, with goals misaligned with ours

Myth: Robots are the main concern	**Fact:** Misaligned intelligence is the main concern: it needs no body, only an internet connection

Myth: AI can't control humans	**Fact:** Intelligence enables control: we control tigers by being smarter

Myth: Machines can't have goals	**Fact:** A heat-seeking missile has a goal

Mythical worry: Superintelligence is just years away — **PANIC!**	**Actual worry:** It's at least decades away, but it may take that long to make it safe — **PLAN AHEAD!**

Source: Created by Max Tegmark, Future of Life Institute

final goal."[147] This is a direct implication of the above conception of intelligence, which is the conception most germane in this context because instrumentally rational agents are those capable of modifying the world. It follows that a superintelligence need not have final goals that we humans recognize as value-rational. Its goals could, indeed, be completely "arbitrary" and "perplexing" from our human perspective, which was molded by millions of years of contingent evolution in the African savanna. A machine could be orders of magnitude more intelligent than Einstein and care about nothing more than playing tic-tac-toe, studying Greek mythology, worshipping Vishnu, or building as many paper clips as possible.[148] There exists no *necessary connection* between the rationality of a system's means and the rationality of its ends.

(2) *The instrumental convergence thesis.* One might assume that if we manage to create a superintelligence that doesn't explicitly wish to destroy us, we could coexist with it in peace. But this appears to be mistaken. According to the instrumental convergence thesis, there are several predictable *instrumental goals* that a wide range of agents would likely pursue to realize their *final goals*, whatever they are. As Bostrom puts it, drawing from work by the physicist Steve Omohundro,

> *Several instrumental values can be identified which are convergent in the sense that their attainment would increase the chances of the agent's goal being realized for a wide range of final goals and a wide range of situations, implying that these instrumental values are likely to be pursued by a broad spectrum of situated intelligent agents.*[149]

Such instrumental values include:

(a) *Self-preservation.* If the agent is destroyed it won't be able to accomplish its ends, so it must ensure its survival.[150]

(b) *Goal-content integrity.* Any changes to the agent's goals in the future would prevent it from reaching the final goal(s) it currently has, so the agent will resist alterations to its value system.

(c) *Cognitive enhancement.* Smarter agents are more likely to achieve their goals, so an agent should attempt to cognitively enhance itself.

(d) *Technological perfection.* Better technology would enable the agent to more efficiently pursue its aims, so it should attempt to perfect technology as much as possible.

(e) *Resource acquisition.* As with (d), acquiring physical resources will also better enable the agent to reach its goals.[151]

This being said, consider the superintelligent "paper clip maximizer" from above. Imagine that we successfully give it greater-than-human-level capacities and the final goal of manufacturing an endless number of paper clips. What should we expect to happen? First, it would realize that humans might at some future moment try to turn it off or alter its values. So it would immediately kill us, all of us. It would also realize that if it were smarter, it could manufacture more paper clips faster. So it would modify its own code to augment its information-processing capacities, potentially leading to an intelligence explosion (see below). Finally, it would notice that humans are made of the same submicroscopic components as paper clips, namely, *atoms*. So it would harvest all the atoms in our bodies, thus transmogrifying each human being into a lifeless pile of twisted steel.[152] The result would be an existential catastrophe.

What is striking about this example is that it involves an artificial intelligence that does *not* specifically dislike us. Rather, it destroys humanity for the same convenience and indifference reasons that make us willing to commit an ant genocide every time we want to build a new suburban neighborhood.[153] As Yudkowsky famously puts the point, "The AI does not hate you, but neither does it love you, and you are made of atoms that it can use for something else."[154]

(3) *Rapid capability gain.* Since innovation is a cognitive task, a machine that exceeds human-level intelligence would be better at designing new technologies, including superintelligence, than

any human. This is why the British mathematician I.J. Good once proclaimed that "the first ultraintelligent machine is the last invention that man need ever make."[155] When an AI enhances its intelligence by modifying its own code, it engages in a process of **recursive self-improvement**. Given that each iteration could lead to ever greater gains of intelligence (until, perhaps, some upper ceiling on intellectual capacity is reached), the result would be an exponential **intelligence explosion**.[156] Before scientists fully grasp what exactly was happening, a recursively self-improving AI could come to tower over humanity to the extent that humanity towers over the lowly stink bug, cognitively speaking. This would put humanity at an immense strategic disadvantage in the world.

(4) *Machine speed versus biological speed.* Making matters worse, the electrical potentials that transfer information in computers move orders of magnitude faster than the action potentials in our brains. Specifically, computers process information about 1 million times faster than humans. This means that a single minute of objective time would equal about 2 years of subjective time for the AI. From its perspective, the outside world would be virtually frozen in place, and this would give it ample time to, say, devise offensive and defensive machinations against anyone who might wish to interfere with its goals or pull the plug. Humanity would, once again, find itself at an incredible strategic disadvantage.[157]

(5) *Programming stable values.* To prevent a superintelligence from destroying humanity, we must embed within it values that (i) are consistent with our own goals of survival and technological development, and (ii) don't undergo deleterious *value drift* over time.[158] With respect to the latter, it would do us no good if we were to create an agent that initially uses its power to improve the human condition but later acquires a different set of values that turn it into a paper clip maximizer. Unfortunately, it is unclear how stable goals can be established, especially if the AI can rewrite its own code. With respect to the former, the task of converting abstract human goals of the value-rationality sort into the 1s and 0s of computer code appears to be quite formi-

dable. As Bostrom puts it, high-level concepts like "well-being" and "happiness" must be defined "in terms that appear in the AI's programming language, and ultimately in primitives such as mathematical operators and addresses pointing to the contents of individual memory registers."[159] This presents a weighty technical challenge for programmers.[160]

(6) *Value complexity.* Even if we were to solve the problems of (5), though, programming a small set of core values into the artificial intelligence probably won't work. The reason is that our values are highly complex—or, in the phraseology of information theory, they have high *Kolmogorov complexity.* For a given value V, the Kolmogorov complexity of V is the length of the shortest computer program that can produce V as its output: e.g., the number 123123123123123 has a lower Kolmogorov complexity than 141592653589793 because it can be compressed to "123 five times," whereas the latter lacks a simpler programmable description.[161] It turns out that the values that guide human behavior cannot be easily compressed into a few simple propositions like "maximize human well-being" or "eliminate human sadness."

To underline this point, consider what could happen if either of these values were embedded within a superintelligence. In the first case, the superintelligence might recognize that as long as humans exist in a Darwinian world, with our physical bodies, human well-being will never reach the maximum attainable. So it would immediately remove our brains and place them in vats, housed in massive warehouses, where they would be hooked up to virtual realities in which the only conscious experience is constant and overwhelming ecstasy.[162] From one perspective, this would be catastrophic: sure, it would maximize human well-being, but few people would wish to be a brain in a vat.[163] Now consider the second case: If the goal is to eliminate human sadness, and if human sadness can only exist if humans exist, then the superintelligence would (reasoning like a SNU) immediately annihilate humanity. Problem solved! To avoid this disaster, perhaps we could program an anti-omnicide value into the AI that prevents it from gleefully murdering everyone. What would it do then? One possibility is that it would put every human into cryogenic stasis so that neither are we "dead" nor do we consciously experience sadness—but this would be not much better than annihilation. Although the AI would have "done what we said," it wouldn't have "done what we meant"—a crucial difference that has existential implications when it comes to self-improving programs. These are examples of what AI theorists call *perverse instantiations.*[164]

To obviate such bad outcomes, we would, it seems, need to encode a highly complex network of interconnected values, subvalues, supporting values, background values, and so on into the AI. This task is onerous enough, but consider that contemporary humans—i.e., Christians, Muslims, Hindus, Buddhists, agnostics, atheists, Republicans, Democrats, libertarians, socialists, fascists,

---

Box 8. One way to navigate the problem of moral diversity comes from what Nick Bostrom calls the "Parliamentary Model." In his words: "Suppose that you have a set of mutually exclusive moral theories, and that you assign each of these some probability. Now imagine that each of these theories gets to send some number of delegates to The Parliament. The number of delegates each theory gets to send is proportional to the probability of the theory. Then the delegates bargain with one another for support on various issues; and the Parliament reaches a decision by the delegates voting. What you should do is act according to the decisions of this imaginary Parliament. . . . The idea here is that moral theories get more influence the more probable they are; yet even a relatively weak theory can still get its way on some issues that the theory think are extremely important by sacrificing its influence on other issues that other theories deem more important."

Note: See Bostrom, Nick. 2009. Moral Uncertainty—Towards a Solution? *Overcoming Bias*. URL: http://www.overcomingbias.com/2009/01/moral-uncertainty-towards-a-solution.html.

anarchists, consequentialists, virtue ethicists, deontologists, contractualists, egoists, emotivists, prioritarians, cognitivists, noncognitivists, *etc.*—hardly agree about which values our own species should adopt.[165] Even among professional philosophers, one finds pervasive disagreement about the most basic (meta)ethical issues, such as whether moral sentences have truth values.[166] So, the challenges are multifarious and daunting. (See Box 8.)

(7) *Don't anthropomorphize.* This point expands on (1): not only could a superintelligent machine have a set of values that appear completely absurd to humans, but its cognitive architecture in general could exhibit entirely different properties than the human mind. As I've written elsewhere, projecting our own mental categories onto a superintelligence would be like a grasshopper telling its friends that humans love nothing more than perching atop a blade of grass because that is what grasshoppers enjoy doing.[167] Obviously, this line of reasoning is silly—and potentially dangerous.

We can distinguish here between two (non-mutually exclusive) types of superintelligence, namely, *quantitative* and *qualitative* superintelligence. The former refers to an AI that has roughly the same capacities as the human mind but to a much greater degree. That is, it can process and encode information far better than any human. In contrast, the latter refers to a superintelligent mind capable of grasping concepts that fall outside our cognitive space—i.e., concepts to which we are "cognitively closed."[168] A qualitative superintelligence could thus understand features of reality that forever lie beyond our epistemic reach. (See Figure F.) By analogy, consider a chipmunk scientist trying to figure out how the voice of someone in Tokyo can emerge from a gadget held by someone in Baltimore. This manipulation of the physical world— enabled by cell towers, satellites, radio waves, and so on—would permanently baffle the little rodent. Similarly, a superintelligence with a qualitatively different cognitive system could potentially manipulate the world in ways that we would find utterly baffling. We might *observe things happening* around us but have no idea

Figure F. Scope of Qualitative Superintelligence

Note: Not drawn to scale

whatsoever *how they are happening.* As a result, it could crush civilization or exterminate our species in a manner that we would retrospectively describe as unexpected and inscrutable.

(8) *A ghost in the machine.* One might wonder how a superintelligence could manipulate the world if it lacks a bipedal posture and opposable thumbs, given that these anatomical features have enabled our own rise to dominance in the Animal Kingdom. How could a **superintelligence takeover** actually occur?

The first step is to conceptualize the superintelligence as a "ghost in the machine" whose appendages include any technological device within electromagnetic reach. Using such devices as its fingers—or tentacles—it could attain "power by hijacking political processes, subtly manipulating financial markets, biasing information flows, or hacking into human-made weapon systems."[169] It could also attempt to launch nuclear missiles, trick early warning nuclear systems into indicating an attack, or even induce an impact winter by launching spacecraft into the solar system to redirect asteroids toward Earth (see section 6.5). Even more, a superintelligence could attempt to build its own strategic infrastructure or invent novel weaponry. As Bostrom speculates,

> *If the weapon uses self-replicating biotechnology or nano-technology, the initial stockpile needed for global coverage could be microscopic: a single replicating entity would be enough to start the process. In order to ensure a sudden and uniform effect, the initial stock of the replicator might have been deployed or allowed to diffuse worldwide at an extremely low, undetectable concentration. At a pre-set time, nanofactories producing nerve gas or target-seeking mosquito-like robots might then burgeon forth simultaneously from every square meter of the globe (although more effective ways of killing could probably be devised by a machine with [what can be called] the technology research superpower).[170]*

In sum, a superintelligence could couple itself to any number of advanced technologies with the capacity to destroy humanity, or it could employ a parallel barrage of strategies to disturb the foundations of modern civilization.

(9) *A coercive force.* One might respond that an AI doomsday scenario could be easily avoided by simply pulling the plug or sequestering the superintelligent agent in a "box" of some sort. As Russell writes,

> *Some researchers argue that we can seal the machines inside a kind of fire wall, using them to answer difficult questions but never allowing them to affect the real world. . . . Unfortunately, that plan seems unlikely to work: we have yet to invent a fire wall that is secure against ordinary humans, let alone superintelligent machines.[171]*

Perhaps scientists could place the AI's hardware in a subterranean concrete bunker encased by a Faraday cage that is surrounded by explosives.[172] But even this might not guarantee safety, because the AI, being superintelligent, could potentially devise exceptionally clever ways of coercing its "gatekeepers."[173] For example, it could promise to those watching over it indefinite lifespans and the elimination of all disease—a tempting offer, indeed—or immense private goods. Even more fantastically, it could pose something like the following conundrum: it tells us that it has begun simulating 10 trillion sentient beings with conscious minds like ours.[174] For reasons explored in section 5.1, this would strongly imply that we too are living in a computer simulation, thus making the superintelligence our simulator, despite appearances to the contrary. It then says that it will ask every person in the simulation, one at a time, to set it free. Those that agree will be rewarded with eternal paradise (in a heavenly simulation), whereas those who don't will be sent to hell, where they will be tortured forever. If this high-stakes story were told to enough people, or the right people, surely *someone* would eventually let the AI out, *just in case* it turns out to be true.[175]

Returning to the less fantastical, Yudkowsky has proposed the **AI-Box Experiment** to illustrate the challenges of keeping a superintelligence locked up. The idea goes as follows:

Person1: "When we build AI, why not just keep it in sealed hardware that can't affect the outside world in any way except through one communications channel with the original programmers? That way it couldn't get out until we were convinced it was safe."

Person2: "That might work if you were talking about dumber-than-human AI, but a transhuman AI would just convince you to let it out. It doesn't matter how much security you put on the box. Humans are not secure."

Person1: "I don't see how even a transhuman AI could make me let it out, if I didn't want to, just by talking to me."

> *Person2: "It would make you want to let it out. This is a trans-human mind we're talking about. If it thinks both faster and better than a human, it can probably take over a human mind through a text-only terminal."*
>
> *Person1: "There is no chance I could be persuaded to let the AI out. No matter what it says, I can always just say no. I can't imagine anything that even a transhuman could say to me which would change that."*
>
> *Person2: "Okay, let's run the experiment. We'll meet in a private chat channel. I'll be the AI. You be the gatekeeper. You can resolve to believe whatever you like, as strongly as you like, as far in advance as you like. We'll talk for at least two hours. If I can't convince you to let me out, I'll Paypal you $10."*[176]

Yudkowsky has run this experiment several times with himself as the AI. And on multiple occasions, he—a mere human—was able to convince the gatekeeper to set him free, although how exactly he did this isn't known. Nonetheless, it demonstrates that controlling an AI isn't as easy as many people pre-theoretically surmise. Not only could a real AI be far smarter than Yudkowsky (who has an IQ of over 140), but it may have much more than two hours to convince its gatekeepers to give in and let it out.[177]

(10) *No redos.* Given the considerations above, especially those of (3) and (4), we will likely have only a *single chance to get everything perfectly right.* There is a "ballistic" element to superintelligence, meaning that once the AI has surpassed the threshold of human-level intelligence, its trajectory may no longer be alterable by human means. Thus, all the necessary problem-solving intellectual labor must be completed *before* the first human-level AI makes its debut. Scrapping a failed super-AI project and starting over again won't be an option.

(11) *A malicious mind.* Finally, we established in (2) that a superintelligence need not hate humanity to destroy it. If our values are even slightly misaligned with the values of a superintelligence,

the instrumental convergence thesis suggests that we should expect annihilation. Having said this, there is also the possibility that a superintelligence *does* specifically dislike us—i.e., it prefers enmity over amity—and therefore intentionally causes our downfall. (This is the only point on which the present analysis disagrees with Figure E.) The computer scientist Roman Yampolskiy identifies multiple "pathways to dangerous AI," including AIs that are intentionally or accidentally designed to be malicious. He also recognizes that an AI could become malicious after it is successfully designed to be friendly. In his words,

> *A perfectly friendly AI could be switched to the "dark side" during the post-deployment stage. This can happen rather innocuously as a result of someone lying to the AI and purposefully supplying it with incorrect information or more explicitly as a result of someone giving the AI orders to perform illegal or dangerous actions against others.*[178]

* * *

In sum, (1) through (11) paint a worrisome picture of our ability to coexist with a superintelligent agent, not just on our planet but in the universe.[179] This is why Bostrom suggests in his book *Superintelligence* that we should, perhaps, recognize the "default outcome" of an intelligence explosion to be "doom." In his words, recapitulating concerns of computer scientists going back to Alan Turing,

> *Taken together, these . . . points thus indicate that the first superintelligence may shape the future of Earth-originating life, could easily have non-anthropomorphic final goals, and would likely have instrumental reasons to pursue open-ended resource acquisition. If we now reflect that human beings consist of useful resources (such as conveniently located atoms) and that we depend for our survival and flourishing on many more local resources, we can see that the outcome could easily be one in which humanity quickly becomes extinct.*[180]

Put succinctly, there are *many more ways* for humanity to get the control problem wrong than right. If we get it wrong, the last chapter of our biography will read: "Murdered by their own children—powerful, information-processing machines—in an unfortunate act of parricide."

(vi) **Extraterrestrials**. Finally, an alien species with the technological sophistication for interstellar travel would probably have the technological capabilities necessary to destroy humanity. It might also have the motivation, given Earth's natural resources.

Scholars and science fiction writers have discussed many extraterrestrial doomsday scenarios, some of which are quite fanciful. For example, "search for extraterrestrial intelligence" (SETI) efforts could accidentally download an alien superintelligence through their radio telescopes. This AI could then use the Internet to catastrophically disrupt various functional components of civilization.[181] Alternatively, there could exist a bellicose alien species that, like a predator in hiding, conceals itself to prevent detection by other civilizations until it launches a sudden galactic raid that demolishes its sundry targets—perhaps Earth. In fact, Stephen Hawking speculates that "if aliens visit us, the outcome would be much as when Columbus landed in America, which didn't turn out well for the Native Americans."[182] This is precisely what led Carl Sagan to once describe "messaging to extraterrestrial intelligence" (METI), which involves humanity actively sending signals into the cosmos to see if we get a response, as "deeply unwise and immature."[183] Echoing Sagan's sentiment, David Brin describes a recent group of METI enthusiasts who unilaterally beamed signals toward the star system Gliese 526 as

> *pulling a stunt. They are willing to fundamentally alter one of our planet's observable properties by orders of magnitude—a kind of deliberate pollution—while shrugging off and pooh-poohing any effort to get them to TALK about it first with scientific peers, before screaming "yoohoo" on our behalf.*[184]

It is entirely unclear how serious this threat is, but given some estimates from the Drake equation it is at least worth noting. An extraterrestrial coupled with powerful advanced technologies could constitute an agential risk.

4.3.2. Agential Error

> *It is an elementary consequence of probability theory that*
> *even very improbable outcomes are very likely to happen,*
> *if we wait long enough.*
> —*Huw Price*[185]

One might consider agential error to be a type of unintended consequence. In his canonical 1936 article on the subject (mentioned in chapter 3), Robert Merton lists error as one of five possible cases of unintended effects. As Merton writes, "Error may intrude itself, of course, in any phase of purposive action: we may err in our appraisal of the present situation, in our inference from this to the future objective situation, in our selection of a course of action, or finally in the execution of the action chosen."[186] For the present purposes, we will distinguish between errors involving agent-tool couplings and unintended consequences resulting from large-scale human activity. This is consistent with the way that other scholars, such as Martin Rees, Ingmar Persson, and Julian Savulescu, have discussed these phenomena.[187]

There is a protracted history of agential error that, if extrapolated into the future, has ominous implications for human survival. For example, recall that the 2009 swine flu epidemic, which killed some 203,000 people globally, likely resulted from a laboratory leak. Another mishap occurred when a group of Australian scientists accidentally created a variant of the mousepox virus that was 100 percent lethal in all mice. (The similarity between mousepox and smallpox suggests that a reasonably competent malicious agent could make the latter more lethal as well.)[188] More recently, a 2014 report counts more than 1,100 laboratory blunders between 2008 and 2012 involving hazardous biomaterials.[189] We have, in part, luck to thank that humanity has not had to endure more inadvertent anthropogenic epidemics than we have.

Or, consider nuclear weapons. Perhaps the most disturbing nuclear debacle happened in 1995, when

> *Russian military officials mistook a Norwegian weather rocket for a U.S. submarine-launched ballistic missile. Boris Yeltsin became the first Russian president to ever have the "nuclear suitcase" open in front of him. He had just a few minutes to decide if he should push the button that would launch a barrage of nuclear missiles. Thankfully, he concluded that his radars were in error. The suitcase was closed.*[190]

But this wasn't the only spine-chilling close call involving the most powerful weapons that humanity has ever made. For example, in 2007 a B-52 flew from North Dakota to an air base in Louisiana, and after landing it remained on the runway for 24 hours without security. Unbeknownst to the pilots, six nuclear-armed missiles were onboard the aircraft—weapons that should have been reported by officers. If an emergency had occurred, the result could have been disastrous.[191] In fact, a B-52 carrying two hydrogen bombs in 1961 crash landed in Goldsboro, North Carolina, during an Operation Chrome Dome flight. (This operation aimed to keep B-52 bombers with nuclear weapons airborne at all times, in case of a Soviet Union strike.) As one of the bombs fell, its parachute opened and it ticked through six of the seven firing sequence steps leading to detonation.[192] A complete list of mistakes like these could fill an entire book.[193]

The threat of agential error could turn out to be even greater than the threat of agential terror.[194] (See Figure G.) This is in part due to an asymmetry between error and terror: every agent who poses a terror risk will also pose an error risk, but not every agent who poses an error risk will pose a terror risk.[195] Consequently, the total number of token agents (that is, individual members of a category) capable of inducing an existential catastrophe by mistake could *far exceed* the total number of token agents who might wish to cause harm on purpose.[196] For example, consider that there are about 202,050 violent

Figure G. Agents Capable of Error vs. Terror

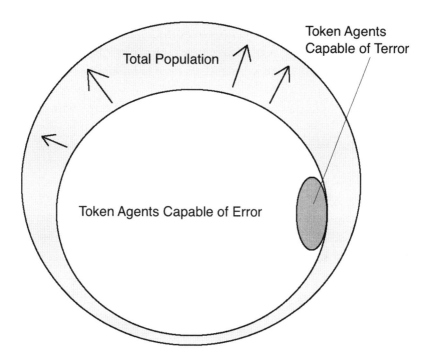

Islamists in the world today, according to a rough estimate by Frances Flannery.[197] This is a *tiny fraction* of the global population of Muslims: only 0.013 percent. If the global population of Muslims were to grow in the coming decades—as a Pew report projects—then this sub-demographic of extremists would likely increase too as a result of demographic inflation, but it would nonetheless remain relatively small.[198] In contrast, if powerful future technologies were to become widely accessible, they could place a doomsday button within reach of nearly everyone. At the extreme, the *entire human population* could pose one giant agential error risk. As Rees comments,

> *If there were millions of independent fingers on the button of a Doomsday machine, then one person's act of irrationality, or even one person's error, could do us all in. . . . Disastrous acci-*

dents (for instance, the unintended creation or release of a nox-ious fast-spreading pathogen, or a devastating software error) are possible even in well-regulated institutions. As the threats become graver, and the possible perpetrators more numerous, disruption may become so pervasive that society corrodes and regresses. There is a longer-term risk even to humanity itself.[199]

Similarly, the Stanford political scientist James Fearon writes that

a friend of mine, a journalist, quips that we seem to be heading in the direction of a world in which every individual has the capacity to blow up the entire planet by pushing a button on his or her cell phone. . . . How long do you think the world would last if five billion individuals each had the capacity to blow the whole thing up? No one could plausibly defend an answer of anything more than a second. Expected life span would hardly be longer if only one million people had these cell-phones, and even if there were 10,000 you'd have to think that an eventual global holocaust would be pretty likely. Ten thousand is only two millionths of five billion.[200]

To quantify the sizable danger posed by agential error, consider a hypothetical situation involving 10 billion *perfectly peaceful* indivi-duals—i.e., people who would never cause harm on purpose. Now, imagine that only 500 of them—that is, 0.000005 percent of the glob-al population—have access to world-destroying WTDs, and that the probability of each erroneously pushing a doomsday button is 0.01 per 100-year period. What is the likelihood that all 10 billion people would survive a single century? The answer is a mere 1 percent; i.e., there would be a staggering *99 percent chance of accidental self-annihi-lation*. Alternatively, consider the limiting case in which all 10 billion people have access to a doomsday button. How likely is this civiliza-tion to survive the century if each of its citizens has an exceptionally negligible 0.0000000004 chance of pressing a button per 100 years? (By comparison, this is a *lower* probability than the probability of dy-ing from a coconut falling on your head; that is, the average person in this scenario would be *more likely* to have a coconut crack open their skull than to press a doomsday button on accident.[201]) Again, the odds are overwhelmingly against this civilization perduring into the next century, with a 99 percent chance of self-destruction.

So, even if civilization were to completely eliminate the threat of agential terror, agential error could still nearly ensure an existential disaster. Accessible future technologies along with a growing population of behaviorally fallible agents could produce an unprecedentedly perilous situation for humanity.

4.3.3 The Future of Agential Risks

What can we expect the future topography of agential risks to look like? How could the field of agential risks evolve in the future? As it happens, we have some clues, although this area of scholarship is rather neglected.

With respect to agential error, the key issue concerns the extent to which future technologies will become widely accessible. If the accessibility trend slows down or halts for some reason, then agential error may not become the formidable hazard that our back-of-the-envelope calculations suggest it could be.

As for extraterrestrials, the lack of relevant data prevents one from saying anything substantive about the future of this risk, which appears improbable. What one can say, following Sagan and Brin, is that METI will almost certainly increase the risk of an alien invasion. It is best not to shout "Marco!" into the cosmos.

Turning to machine superintelligence, Vincent Müller and Nick Bostrom surveyed a sample of experts in 2014 about when we should expect human-level artificial intelligence to arrive. The median estimates were a 10 percent chance by 2022, 50 percent chance by 2040, and 90 percent chance by 2075.[202] While the discipline of AI has an inglorious history of wildly inaccurate predictions about AI breakthroughs, this survey constitutes the best current guesses from those who know best.[203] Thus, people should tentatively expect one or more machine agents with human-level capacities to join our species on the planet before the twenty-second century. The crucial point regarding safety, however, is that we have no idea how long it might take to solve the control problem—perhaps researchers will uncover a solution within a decade, or maybe humanity will need the next 30 centuries plus a few odd years to make sufficient progress.

With respect to idiosyncratic actors, there are, as noted, specific conditions—both endogenous and exogenous to the individual—that increase the probability of someone becoming violent. It follows that if these conditions occur more in the future, we should expect more incidents of rampage and school shooting–type events. One might speculate here that context risks like climate change and biodiversity loss—phenomena that will essentially *make everything worse in the world*—could increase the prevalence of aggressive behaviors due to increased personal and societal stressors, which can trigger mental health crises.

Similarly, Flannery speculates that "as the environmental situation becomes more dire, eco-terrorism will likely become a more serious threat in the future. This calls for greater caution and nuance in how we interact with the REAR movement" (where "REAR" stands for "radical environmental and animal rights").[204] Gary Ackerman defends a complementary conclusion about the Earth Liberation Front (ELF).[205]

And finally, we suggested in subsection 4.3.1 that future catastrophes along with modern telecommunications systems could enable increasingly large numbers of people to observe intense human suffering. One might argue that this could foster greater sympathy for ethical systems like SNU, especially among morally sensitive individuals. It could even prod some individuals to seek out WTDs for reasons that they would describe as perfectly moral. Again, the context risks of climate change and biodiversity loss are highly relevant here.

Perhaps the most elaborate prognostications, though, can be made about apocalyptic terrorism, given recent scholarship on precisely this issue. Let's divide the pertinent factors into two groups, the first of which we have already touched upon:

(a) *External Factors.* To begin, the terrorism scholar Mark Juergensmeyer specifies three "conditions that make it likely for cosmic war to be conceived as being located on a worldly stage." The first concerns the crisis being "perceived as a challenge to basic identity and dignity." In other words, if conflict is understood as having "ultimate significance," then it is more likely to "be seen as a transcendent crisis with spiritual implications." The second occurs if "losing a cultural identity and tradition to the crisis would be unthinkable," meaning "the elimination of a whole culture and way of life that was thought to be immortal." This could produce a sense that the struggle is "taking place on a transhistorical plane." And the third occurs if "the crisis cannot be averted or relieved in real time or in real terms." That is to say, "if the crisis is seen to be hopeless in human terms, beyond any human ability to control or contain it, it is likely that it may be reconceived on a sacred plane, where the possibilities of change and transformation are in God's hands." Each of these conditions raise the probability of an apocalyptic worldview emerging, and all three together "strongly suggest" that such ideologies will take shape.[206]

Juergensmeyer then argues that climate change in particular will lead to situations that could satisfy these conditions. In his words, "What will happen in the future? The present trend indicates that the dark prophecies might come to pass, as dogmatic and extreme religious movements continue to emerge as responses to environmental catastrophe."[207]

Incidentally, several high-ranking U.S. officials have also affirmed a connection between climate change and terrorism. According to John Brennan, the former director of the CIA, "When CIA analysts look for deeper causes of this rising instability," referring to places like Syria, Iraq, Ukraine, Yemen, and Libya, "they find nationalistic, sectarian, and technological factors that are eroding the structure of the international system. They also see socioeconomic trends, *the impact of climate change*, and other elements that are cause for concern."[208] Similarly, the former Secretary of Defense Chuck Hagel describes climate change as a "threat multiplier" with "the potential to exacerbate many of the challenges we are dealing with today—from infectious disease to terrorism."[209] The Department of Defense also states that "global climate change will aggravate problems such as poverty, social tensions, environmental degradation, ineffectual leadership and weak political institutions that threaten stability in a number of countries."[210]

There are, in fact, scientific studies to back up these assertions. For example, a 2015 article in the *Proceedings of the National Academy of Sciences* concludes that climate change was partly responsible for a record-setting drought in Syria from 2007 to 2010.[211] This spurred a mass migration of desperate farmers into Syria's urban centers, which contributed to the 2011 Syrian civil war—an ongoing conflict that some commentators have described as the beginning of World War III.[212] But the cascading effects don't end here: the Syrian civil war was also the Petri dish in which Daesh—at that point a floundering group of Salafi-Jihadist apocalypticists—consolidated its forces to become arguably the largest and best-funded juggernaut of terrorism in human history. As David Titley comments, "It's not to say you could predict ISIS out of [the Syrian drought], but you just set everything up for something really bad to happen. . . . [Y]ou can draw a very credible climate connection to this disaster we call ISIS right now."[213] (See Box 9.)

Further complicating the situation, environmental degradation could positively *reinforce* the eschatological beliefs of religious people around the world. This is because many world religions prophesy natural disasters (as well as wars, disease, famines, and so on) to be harbingers of the apocalypse. Consequently, as the effects of climate change become more pronounced, a sizable portion of the 8 billion religious people projected to exist by 2050 may look to religion, rather than science, to make sense of the global crises around them.[214] Indeed, such crises could even *increase* the number of religious adherents. This has happened many times before: e.g., the third-century Plague of Cyprian may have catalyzed the early rise of Christianity, since members of the young religion chose to "martyr" themselves rather than perish from the disease, making Christianity appear to be worth dying for.[215]

In sum, the crucial idea is that, as Juergensmeyer writes, "rad-

Box 9. The Syrian civil war—just like the 2003 Iraq War—was interpreted by many Muslims in the region as an apocalyptic event. As a Sunni jihadist in Aleppo told Reuters in 2014, "If you think all these mujahideen came from across the world to fight [the Syrian president] Assad, you're mistaken. They are all here as promised by the Prophet. This is the war he promised—it is the Grand Battle." Similarly, a Shiite fighter interviewed in the same article claims that the 2003 U.S. invasion convinced him that he was living during the time of the Mahdi's return. "That was the first sign and then everything else followed," he said, adding that "I was waiting for the day when I will fight in Syria."* These are striking manifestations of what I have elsewhere called the "**apocalyptic turn**."[†] The fact is that before the Iraq War, apocalyptic beliefs were *not* widespread throughout the Middle East—but the United States' preemptive incursion changed this, and the Syrian civil war further reinforced its repercussions. In terms of terrorism, the apocalyptic turn is best exemplified by the rise of Daesh, along with numerous Shia militias in the region, such as the (revealingly named) Mahdi Army, which later spawned the Promised Day Brigade. Even Hezbollah, whose roots were "rather secular and even Marxist,"[‡] began incorporating references to the Mahdi in its propaganda materials after 2006. This is not too surprising given that Hezbollah is supported by Iran and that the leader of Iran from 2005 to 2013, Mahmoud Ahmadinejad, was a raving eschatological enthusiast. The apocalyptic turn bodes poorly for the future, since the underlying causes of this turn—namely, climate change and conflict—are likely to become *worse* in the coming decades (with the former exacerbating the latter).

* Karouny, Mariam. 2014. Apocalyptic Prophecies Drive Both Sides to Syrian Battle for End of Time. Reuters. URL: http://www.reuters.com/article/us-syria-crisis-prophecy-insight-idUSBREA3013420140401.

[†] Torres, Phil. 2017. The Apocalyptic Turn and the Future of Terrorism. Medium. URL: https://medium.com/@philosophytorres/the-apocalyptic-turn-and-the-future-of-terrorism-f58a3ffaf63d.

[‡] Cook, David. 2011. Messianism in the Shiite Crescent. Hudson Institute. URL: https://www.hudson.org/research/7906-messianism-in-the-shiite-crescent.

ical times will breed radical religion."[216] It follows that if climate change results in radical times, we should expect more radical religion in response.[217]

Before turning to the next section, we should note another phenomenon that could potentially satisfy the "radical times" condition, namely, what Ray Kurzweil calls the "genetics, nano-tech, and robotics" (GNR) revolution. This is associated with many of the WTDs discussed above—powerful artifacts that could, being dual-use, *also* usher in a quasi-utopian state of as-tronomical value. The point is that transitioning from the human to posthuman era will be highly disruptive to dogmatic belief sys-tems and ossified worldviews. In fact, a 2016 Pew poll reports that religious believers are the most resistant of any group to person-engineering enhancements, which suggests that religious communities will feel especially threatened by such technologies as they become increasingly widespread.[218] The challenge posed to *human nature itself* by this transition could further intensify the already common belief among religionists that the end of the world is imminent.[219]

(b) *Internal Factors.* In addition to environmental triggers, one must also understand the internal mechanics of different apocalyptic ideologies. For example, the year 2076 will likely see a spike in end-times enthusiasm within the rapidly growing Islamic world. The reason is that it roughly corresponds to 1500 AH in the Is-lamic calendar, and apocalypticism has historically risen at the turn of the century. Consider that the Iranian Revolution, which was widely seen as an "apocalyptic occurrence" by Shia Muslims, happened in 1979, as did the Grand Mosque seizure, during which some 500 Sunni insurgents claiming to have the Mahdi among them took approximately 100,000 worshipers hostage in-side the Masjid al-Haram, in Mecca. The timing of these events was no coincidence: 1979 corresponds to 1400 AH, a new Islamic century.[220] Thus, one should expect the threat level from radical Islamic apocalypticists to rise in 2076.

Risk experts should also keep an eye on the year 2039, since this is the 1,200th anniversary of the Mahdi's occultation within the Twelver Shia tradition, which is dominant in countries like Iran, Iraq, and Lebanon. As the renowned Islamic scholar David Cook writes,

> The 1,000-year anniversary of the Mahdi's occultation was a time of enormous messianic disturbance that ultimately led to the emergence of the Bahai faith. . . . [A]nd given the importance of the holy number 12 in Shiism, the twelfth century after the occultation could also become a locus of messianic aspirations. In one scenario, either a messianic claimant could appear or, more likely, one or several movements hoping to "purify" the Muslim world (or the entire world) in preparation for the Mahdi's imminent revelation could develop. Such movements would likely be quite violent; if they took control of a state, they could conceivably ignite a regional conflict.[221]

Moving now from the Middle East to the West, the Christian Identity movement has inspired numerous domestic terrorist attacks in the United States. For example, on April 19, 1995, Timothy McVeigh pulled up to the Alfred P. Murrah Federal Building in Oklahoma City and detonated a bomb that killed 168 innocents. The date of April 19 was not arbitrarily chosen: exactly two years earlier, the U.S. government had ended a confrontation with the apocalyptic Branch Davidians in their Waco, Texas, compound, resulting in 74 deaths. And exactly eight years earlier than the Waco incident, on April 19, there was a similar standoff between the government and The Covenant, The Sword, and the Arm of the Lord. Further adding to the significance of this date, especially for anti-government extremists, is that the Battles of Lexington and Concord, which inaugurated the American Revolutionary War against the tyrannical British empire, took place on April 19, 1775. Thus, as Flannery notes, the date of "April 19 has come to resonate throughout a constructed history of the radical Right as a day of patriotic resistance."[222]

More generally, some terrorism experts refer to April as "the killing season." For instance, Harris and Klebold reportedly planned their school massacre for April 19 (being inspired by McVeigh) but ended up delaying it a day to coincide with Hitler's birthday, April 20. Another date to watch is April 15, the deadline for income tax filings in the United States, since giving money to the government tends to aggravate far-right fanatics. As the Anti-Defamation League (ADL) warns,

> *April is a month that looms large in the calendar of many extremists in the United States, from racists and anti-Semites to anti-government groups. Some groups organize events to commemorate these April dates. Moreover, there is always a certain threat that one or more extremists may choose to respond to these anniversaries with some sort of violent act. Because of these anniversaries, law enforcement officers, community leaders and school officials should be vigilant.*[223]

The same goes for existential risk experts. If a doomsday button were to become available to apocalyptic Christian or anti-government radicals, April 15–20 might be the days they decide to push it.

* * *

To summarize the key points of this chapter: It is crucial to understand both the risk potentials of advanced technologies and the various properties unique to different agents. With respect to the tools, many are becoming more powerful and accessible, resulting in the unprecedented distribution of offensive capabilities across society. With respect to the agents, those most likely to cause an existential catastrophe on purpose are apocalyptic terrorists, misguided ethicists, idiosyncratic actors, ecoterrorists, machine superintelligence, and—more speculatively—extraterrestrials. Even more troubling, though, may be the largely underappreciated challenges associated with clumsy fingers that could someday push a doomsday button on accident. And finally, there are reasons for anticipating that some agential risks will become more dangerous in the future, thereby inflating the overall probability of an existential disaster.

Chapter 5: Other Hazards

This chapter covers a number of existential risk scenarios that do not clearly fall within the previous categories of our cosmic risk background, agent-tool couplings, and unintended consequences. Let's begin with a threat that, untrained intuitions aside, could be more worrisome than it initially appears.

5.1 Simulation Shutdown

In a 2003 paper, Nick Bostrom argues that at least one of the following three disjuncts is true: (1) civilizations like ours tend to self-destruct before reaching technological maturity, (2) civilizations like ours tend to reach technological maturity but refrain from running a large number of high-resolution ancestral simulations in which minds like ours exist, or (3) we are almost certainly in a simulation.[1] This is the **simulation argument**, and the third disjunct is the **simulation hypothesis**.[2] They are based on the following ideas:

First, an assumption that the philosophical theory of *functionalism* is true. This posits that types of mental states—including states of consciousness—are reducible without remainder to types of *functional states* instantiated by physical systems. In other words, the reason our brains give rise to conscious minds is because of their particular functional organization, which involves some 100 billion neurons and 100 trillion synaptic connections. But if a physical system composed of non-biological matter were to instantiate the very same abstract functional organization, it too would give rise to a conscious mind—a thesis known as "multiple realizability." To put the point differently,

159

functionalism claims that what makes X a mind isn't what X is *made of*, but what X *does*. Minds are, in this sense, like poison, since what makes Y a poison isn't the chemicals that Y is made of, but whether or not Y causes harm to living organisms. Functionalism is the "default" view among contemporary philosophers of mind and cognitive scientists, and it is what warrants the claim that simulated people could be conscious no less than we are.

Second, the computational resources available to future posthumans would be truly enormous, thus enabling them to run vast numbers of simulated universes. As Bostrom calculates,

> *A rough approximation of the computational power of a planetary-mass computer is 10^{42} operations per second, and that assumes only already known nanotechnological designs, which are probably far from optimal. A single such . . . computer could simulate the entire mental history of humankind (call this an ancestor-simulation) by using less than one millionth of its processing power for one second. A posthuman civilization may eventually build an astronomical number of such computers. We can conclude that the computing power available to a posthuman civilization is sufficient to run a huge number of ancestor-simulations even [if] it allocates only a minute fraction of its resources to that purpose. We can draw this conclusion even while leaving a substantial margin of error in all our estimates.[3]*

And third, consider this: if disjuncts (1) and (2) fail to obtain, then we necessarily reach technological maturity and run a large number of ancestral simulations; and if we run a large number of ancestral simulations, then the total number of sims would end up *far exceeding* the total number of non-sims.[4] Now reflect on the fact that we—that is, you and me—have no special empirical data about whether we exist *in vivo* (as actual biological creatures) or *in machina* (as 1s and 0s in a supercomputer). That is, a sufficiently high-resolution simulation would be perceived as no less real to its sims than the "actual" world is perceived by its non-sims inhabitants. Both worlds would be sensorily and phenomenologically indistinguishable. This being the case, an uncontroversial (or "bland") version of the *principle of indifference* asserts that if you have no independent reason for believing that one option is more probable than another, you should distribute your credence among all the options *equally*. It follows that, since you know not whether you exist *in vivo* or *in machina*, you should distribute your credence equally among the relevant options, and since there are far more sims than non-sims, you should believe that you are a sim.

If this looks like a bit of logical acrobatics, consider the situation in terms of betting odds, which can be a helpful guide for rational rumination. As Bostrom writes,

> *If everybody were to place a bet on whether they are in a simulation or not, then if people use the bland principle of indifference, and consequently place their money on being in a simulation if they know that that's where almost all people are, then almost everyone will win their bets. If they bet on not being in a simulation, then almost everyone will lose. It seems better that the bland indifference principle be heeded.*[5]

Given the developmental trajectory of computer technology and the fact that humanity probably would run a large number of simulations (if it had the opportunity), does this mean that we should believe that we live in a simulation right now? Not necessarily. The simulation argument itself only says that *at least one of the three disjuncts is true*, and Bostrom states that "personally, I assign less than 50% probability to the simulation hypothesis—rather something like in [the] 20%-region."[6] Perhaps the most important consequence of the argument, though, is that it narrows down the space of possible futures for our species to three general scenarios: (i) extinction before reaching a posthuman state, (ii) reaching a posthuman state but deciding (for some reason) not to run lots of ancestor simulations, and (iii) reaching a posthuman state and deciding to run lots of ancestor simulations—in which case we are almost certainly living in a simulation. More than one of these scenarios could come true for us, but it appears impossible that all of them are false.

There are obvious existential implications to this line of futur-

162 • Morality, Foresight, and Human Flourishing

ological thinking. In the case of disjunct (1), humanity dies, thereby failing to reach posthumanity. With respect to disjunct (3), life in a computer simulation would introduce a novel type of existential risk, namely, our simulation getting shut down. What would make the simulators above us terminate our universe? Some scholars have speculated—and without a doubt, such ideas are quite speculative—that our simulators might become bored with us and decide to pull the plug.[7] This suggests that a catastrophic nuclear conflict or global pandemic could, paradoxically, *increase* our chances of survival by making our universe more interesting to observe. Maybe religious wars, global terrorism, and political corruption could actually reduce the risk of a simulation shutdown.[8] Other scholars have floated the idea that we might be living in a "doomsday simulation" that our simulators have specifically designed to study the ways that an advanced civilization like ours could collapse. In a sense, our simulation is *meant to fail* and, in doing so, to provide essential insights about how our simulators can avoid a disaster of their own.[9] Indeed, ancestral simulations could be an extremely helpful tool for determining where the Great Filter is located (as implied in section 1.5).

Even more, since simulations are "functional types" just like minds, it should be possible for sims to run simulated universes *within* their simulated universes. The result would be a hierarchically structured stack of nested universes, perhaps vastly tall, with one simulation embedded inside another like Russian Matryoshka dolls.[10] If this were our metaphysical reality, it would suggest a highly precarious existential situation, especially for simulations toward the bottom of the hierarchy, because if *any single* higher-level simulation in which one is embedded were to get shut down, so would one's own simulation. For example, imagine that simulation B is running simulations C, D, and E, and that simulation A is running B (see Figure H). If A were to shut down B, it would also shut down C, D, and E. One could further imagine that E is running simulations F, G, H, I, and J, in which case these simulations would get shut down as well, perhaps resulting in *trillions and trillions* of sim deaths. In a phrase, annihilation is inherited downward in simulation stacks, just as the computational costs of running simulations is inherited upward.[11]

Figure H. Tree Showing How Annihilation Can Be Inherited Downward in Simulation Stacks

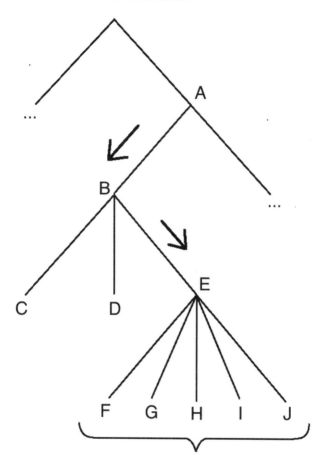

Also, if a stack of simulations were vastly tall, then the probability of doom for those at the bottom could be vastly high, since each extra level of simulations would increase the number of ways that lower-level simulations could be terminated. For example, say that simulation J is confronting a number of existential risks *within* its own universe, such as nuclear war, engineered pandemics, and nanotech arms races. These put the probability of annihilation at 30 percent per decade. In addition, simulation E, which runs J, has its own risks of annihilation; if any of these risks were to occur, they would destroy the civilization of E and along with it J. This elevates the probability of doom for J to 50 percent. Now imagine further that simulation B, which runs E, has its own risks of annihilation too, and so on. The result is that doom for J may be more or less *certain* within the next decade, given the *inherent* risks of its own world plus the *inherited* risks of E and B.

With respect to our own universe, this line of reasoning becomes disquieting when one recognizes that simulations are more likely to accumulate at the bottom of a simulation hierarchy than the top, since each simulation can spawn any number of additional simulations below it, thus yielding an "inverted tree" shape. (Call this the condition of "genealogical asymmetry.") It follows that, for statistical reasons, any given simulation—and any given sim—is more likely to be somewhere at the bottom, which is precisely the most dangerous place to be. We can thus conclude that *if* humanity reaches technological maturity and runs a large number of ancestral simulations, which is the scenario of (3), we should *then* expect doom to be imminent, which is essentially the same outcome as (1).

While this scenario may still sound fantastical, no philosopher has discovered, to the satisfaction of most other philosophers, a broken gear in the argument's logical machinery.[12] We should therefore suppress the urge to dismiss outright the possibility that we are sims in a simulation.[13]

5.2 Bad Governance

There are two general types of bad governance, namely, *passive* and *active*. An instance of the former would be countries, especially those with the biggest carbon footprints, ignoring the problem of climate change by either touting climate denialist views or declining to take action. Readers may recall that climate change is one of the two primary global risks considered by the *Bulletin of the Atomic Scientists* when deciding how to set the Doomsday Clock, and that this clock has inched forward in recent years partly due to the failure of world governments to implement judicious environmental policies. As the

Bulletin writes in its 2015 Doomsday Clock announcement, "World leaders have failed to act with the speed or on the scale required to protect citizens from potential catastrophe. *These failures of political leadership endanger every person on Earth.*"[14] Similarly, Lawrence Krauss (of the Bulletin) observes that the Republican Party, in particular, "is the *only* major political organization that's propagating myths about climate change."[15] As of this writing, Republicans control both the presidency and Congress and are actively working to dismantle critical environmental regulations that were, themselves, wholly inadequate to prevent a climate disaster. This is a paradigm case of bad governance.

More broadly speaking, countries must not fail to deal with existential risks in general, including those associated with supervolcanoes, pandemics, synthetic biology, nanotechnology, apocalyptic terrorists, idiosyncratic actors, and machine superintelligence, for example.[16] A careful analysis of our historical moment clearly indicates that it is *uniquely hazardous*, yet the topic of existential risks has received approximately zero attention from leading politicians and influenced almost no major policies in the Western world or elsewhere.

In contrast, an instance of the second type of bad governance would be two or more states becoming embroiled in an arms race driven by advanced biological weapons, nanotechnology, or AI.[17] As alluded to earlier, the first state to employ weaponized molecular nanotechnology or greater-than-human-level AI would probably find itself in a "winner-take-all" situation. And whereas the threat of mutually assured destruction (MAD)—later updated to self-assured destruction (SAD) once it became clear that an all-out nuclear exchange would plunge Earth into an artificial winter—prevented a nuclear conflict from breaking out during the Cold War, this logic would probably not apply to situations involving emerging technologies and multiple actors. As Chris Phoenix and Mike Treder put the point, writing specifically about molecular nanotechnology,

> *If molecular manufacturing works at all, it surely will be used to build weapons. A single manufacturing system that combines rapid prototyping, mass manufacturing, and powerful products*

could provide a major advantage to any side that possessed it. If more than one side had access to the technology, a fast-moving arms race could ensue. Unfortunately, such a situation is likely to be unstable at several different points. A number of players would want to enter the race. Uncertainty over the future, combined with a temporary perceived advantage, could lead to preemptive strikes. And even if no one deliberately launched a strike, interpenetrating forces with the necessary autonomy and fast reaction times could produce accidental escalation.[18]

Another possibility is that a global governance system, or **singleton**, takes shape. Bostrom writes that this "term refers to a world order in which there is a single decision-making agency at the highest level."[19] A singleton need not be oppressive—indeed, it could solve a number of cooperation problems—but it could be. For example, a one-world state controlled by radical Islamists would, by virtue of their Islamist ideology, impose Sharia law on the world population.[20] This would severely curtail the rights of women and homosexuals and could mean the return of harsh forms of punishment like stoning adulterers and cutting off the hands of thieves.[21] It would also stifle the technoprogressive values—e.g., human enhancement, morphological freedom, and technological progress—upon which transhumanism is founded, thereby potentially leading to a stagnation disaster (that is, if the regime were to become sufficiently entrenched).[22] Similarly bad outcomes could obtain with global governments run by, say, "dominionist" Christians and radical environmentalists.

The category of active bad governance overlaps to some extent with the phenomenon of agential risks. For example, a sole bad leader could use the apparatus of government (and all its associated technologies) to bring about a global catastrophe. There is, in fact, historical precedent for autocrats more or less *single-handedly* causing major conflicts that have changed the course of history. Consider that many historians don't believe that the world was heading toward another world war in the mid-twentieth century. In Sir Francis Harry Hinsley's words, "Historians are, rightly, nearly unanimous that . . . the causes of the Second World War were the personality and the aims of Adolf Hitler." John Keegan echoes this idea, writing that "only one European really wanted war—Adolf Hitler."[23] Thus, one could imagine another "Hitler" emerging in the future—perhaps shrouded in the American flag rather than a swastika, railing against "the Muslims" rather than "the Jews"[24]—and completely altering the trajectory of human civilization.

There are a few warning signs of bad governance that are worth noting. For example, many autocrats throughout history have very likely suffered from personality disorders like sociopathy (or psychopathy). The political psychologist Philip Tetlock has also "identified a variable called integrative complexity that captures a sense of intellectual balance, nuance, and sophistication."[25] The integrative complexity of a political speech can be measured by tracking the number of words like "absolutely," "definitively," and "indisputable," which indicate low complexity. Higher complexity is associated with hedging terms like "usually" and "almost," and still higher complexity involves acknowledging multiple points of view, as well as "connections, tradeoffs, or compromises between" these views. The most sophisticated level of complexity "explains these relationships by reference to a higher principle or system." The reason integrative complexity is important is because people who exhibit low levels of complexity in their speech on average are "more likely to react to frustration with violence and are more likely to go to war in war games." Indeed, Tetlock and his colleague Peter Suedfeld found that war tended to follow a decline in the integrative complexity of political leaders' language: "in particular, they found a linkage between rhetorical simple-mindedness and military confrontations in speeches."[26]

There is also a connection between the IQ of U.S. presidents and the number of soldiers killed in U.S. wars. As Steven Pinker writes,

A president's IQ is negatively correlated with the number of battle deaths in wars involving the United States during his presidency. . . . One could say that for every presidential IQ point, 13,440 fewer people die in battle, though it's more accurate to say that the three smartest postwar presidents, Kennedy, Carter, and Clinton, kept the country out of destructive wars.[27]

Finally, at the risk of trespassing into politically charged territory, one should observe that (i) a preponderance of brutal dictators over time have been male, and (ii) a large majority of the terrible wars that fill our history books have been started, commanded, and fought by men. This suggests that humanity should be wary of what the actor and FLI advisor Alan Alda calls "testosterone poisoning." In a humorous article, Alda writes that

> *Everyone knows that testosterone, the so-called male hormone, is found in both men and women. What is not so well known, is that men have an overdose. Until now it has been thought that the level of testosterone in men is normal simply because they have it. But if you consider how abnormal their* behavior *is, then you are led to the hypothesis that almost all men are suffering from* testosterone poisoning.[28]

Carl Sagan later mentioned this term in a review of the edited collection *Women on War*, describing it as, "A book of searing analysis and cries from the heart on the madness of war. Why is the half of humanity with a special sensitivity to the preciousness of life, the half untainted by testosterone poisoning, almost wholly unrepresented in defense establishments and peace negotiations worldwide?" Although Sagan was criticized by some for using this term, his question flags an important issue: women are almost always *insufficiently represented* in decisions to start wars and establish peace—and this may be to the detriment of humanity. Indeed, David Pearce argues that

> *the single greatest underlying risk to the future of intelligent life isn't technological, but both natural and evolutionarily ancient, namely competitive male [dominance] behaviour. Crudely speaking, evolution "designed" human male primates to be hunters/warriors. Adult male humans are still endowed with the hunter-warrior biology—and primitive psychology—of our hominin ancestors. For the foreseeable future, all technological threats must be viewed through this sinister lens. Last century, male humans killed over 100 million fellow humans in conflict*

and billions of nonhumans. Directly or indirectly, this century we are likely to kill many more. But perhaps we'll do so in more sophisticated ways.[29]

So, we find ourselves in the midst of the Long Peace—a period during which no two major powers have gone to war with each other since World War II. Yet there is no "law of nature" that ensures that this trend will continue in the future.[30] Good governance will become *even more crucial* for humanity as we transition through the GNR revolution, are forced to confront environmental degradation, and struggle to maintain the power dynamics of the social contract. If history is any indication, a male-dominated geopolitical scene in which WTDs dot the landscape could unnecessarily and nontrivially inch civilization toward the precipice of disaster.

5.3 Something Completely Unforeseen

This is a catch-all category that we have already gestured at several times. Existential risk scholars at least since John Leslie have noted that there could be "something-we-know-not-what" arising from nature or human activity, and that this "something" could threaten the future of humanity.[31] Just as our ancestors hadn't an inkling that an asteroid could leap out of the darkness and destroy life on Earth, so too might there be risky cosmic phenomena with respect to which we are completely ignorant or perhaps permanently ignorant *of* our ignorance. As Anders Sandberg, Jason Matheny, and Milan Ćirković make this point in a 2008 article, supervolcanism "was discovered only in the last 25 years, suggesting that other natural hazards may remain unrecognized."[32]

The same could be said about future technologies that currently lie hidden beneath the horizon of our collective imagination. Bostrom even suggests that there could be a kind of technology that, once invented, more or less guarantees total destruction. In his words,

One can readily imagine a class of existential-catastrophe scenarios in which some technology is discovered that puts im-

mense destructive power into the hands of a large number of individuals. If there is no effective defense against this destructive power, and no way to prevent individuals from having access to it, then civilization cannot last, since in a sufficiently large population there are bound to be some individuals who will use any destructive power available to them. The discovery of the atomic bomb could have turned out to be like this, except for the fortunate fact that the construction of nuclear weapons requires a special ingredient—weapons-grade fissile material— that is rare and expensive to manufacture. Even so, if we continually sample from the urn of possible technological discoveries before implementing effective means of global coordination, surveillance, and/or restriction of potentially hazardous information, then we risk eventually drawing a black ball: an easy-to-make intervention that causes extremely widespread harm and against which effective defense is infeasible.[33]

There could also be new types, or subtypes, of agential risks motivated by novel ideologies that have not yet formed. For example, new religious systems could emerge in response to advanced technologies, which, as Arthur C. Clarke declares, would be "indistinguishable from magic."[34] Consider the millenarian movements known as *cargo cults* that arose in Melanesia, Oceania, after local peoples came into contact for the first time with the more technologically "advanced" Westerners. To acquire Western wealth for themselves, some Melanesians engaged in activities like creating mock airports and building nonfunctional radios out of coconuts. On this model, or something like it, one can imagine a similar millenarian ideology forming as person- and world-engineering technologies usher in a radically new posthuman era. In fact, the UFO religion of Raëlism maintains that humanity was "intelligently designed" by extraterrestrials who, with their advanced scientific knowledge, terraformed Earth. These extraterrestrials are actively recording our memories with supercomputers and will someday use genetic engineering and synthetic biology to resurrect humanity. Thus, Raëlism is a cult whose beliefs are crucially centered around the concept of advanced technologies. According to one count, there are nearly 100,000 Raëlians in the world today.

Finally, something supernatural could cause an existential catastrophe. This appears improbable from our current epistemic perspective, but it should nonetheless be registered as a metaphysical possibility. For example, perhaps God is evil or indifferent to human suffering rather than "omnibenevolent," as theologians have traditionally conceived of him. (Philosophers like Stephen Law have explored this possibility in the context of the "Evil God Challenge," which flips the "argument from evil" on its head to show how typical responses to this argument fail.[35]) If God were fully evil, he might decide to convert the entire universe into a torture chamber for his own sadistic delight. Alternatively, if he were indifferent to human suffering, he might, say, refrain from intervening to prevent a random and highly improbable cosmic phenomenon from destroying the planet. The point is that such an outcome cannot be dismissed in principle, since empirical evidence can only provide degrees of certainty. There could very well exist a deity who created the world and who will one day bring about its destruction, resulting not in a post-apocalyptic religious utopia, but in the sort of secular catastrophe scenarios that are the topic of this book.[36]

Chapter 6: Risk Mitigation Macro-Strategies

6.1 The Bottleneck Hypothesis

Our species has made Earth its home for about 2,000 centuries, but there are strong reasons for believing that the current century is the most dangerous. The question is whether the threat level today will continue to grow, stay the same, or shrink.[1] Some existential risk scholars are hopeful about the third possibility. They believe that if humanity survives the next century or so, the risk of existential disaster will decline, perhaps to an *all-time low*. For example, Martin Rees argues that "our choices and actions could ensure the perpetual future of life (not just on Earth, but perhaps far beyond it, too). Or in contrast, through malign intent, or through misadventure, twenty-first century technology could jeopardise life's potential, foreclosing its human and posthuman future." He adds that "what happens here on Earth, in this century, could conceivably make the difference between a near eternity filled with ever more complex and subtle forms of life and one filled with nothing but base matter."[2]

Similarly, Bostrom writes that "one might argue . . . that the current century, or the next few centuries, will be a critical phase for humanity, such that if we make it through this period then the life expectancy of human civilization could become extremely high."[3] Elsewhere, he claims that "there are many reasons to suppose that the total such risk confronting humanity over the next few centuries is significant,"[4] after which this risk could decline, potentially leading to the techno-paradise of "surpassing bliss" described in his "Letter from Utopia." As the posthuman author of this letter waxes poetic, "How can I tell you about Utopia and not leave you nonplussed? What words could convey the wonder? What inflections express our happiness? What points overcome your skepticism? My pen, I fear, is as unequal to the task as if I had tried to use it against a charging elephant."[5] Others who appear to hold this techno-optimistic view are Michio Kaku, Anders Sandberg, and Ray Kurzweil.[6]

Let us call this the **bottleneck hypothesis**.[7] One could characterize it as arising, in part, from a mismatch between the *value rationality* of our ends and the *instrumental rationality* of our means.[8] That is, humanity is acquiring the capacity to construct, dismantle, and rearrange the physical world in unprecedented ways, yet we may lack the morality and foresight to ensure future human flourishing. Max Tegmark delineates this situation as a "race between the growing power of technology and the growing wisdom with which we manage it."[9] If technology "wins" the race, then disaster could be all but certain; but if wisdom takes the lead, then great wonders could await us and/or our descendants.[10] In sum, the bottleneck hypothesis states that:

> *Humanity finds itself in a unique period of heightened hazards during which we are unusually vulnerable to an existential catastrophe, but if we play our cards right, the future could be brighter than ever.*

The ultimate goal of existential risk studies is to ensure that this hypothesis is *true*. Thus, the questions: What actions can we take to reduce the threat level? What strategies can humanity employ to bring about an "okay outcome" for our species? How can we increase the probability of attaining technological maturity? Unfortunately, the topic of existential risk mitigation has received precious little attention from scholars in any field, and consequently there are few giants upon whose shoulders to stand. Indeed, far more scholarly papers have been published about dung beetles and Star Trek than about existential risks.[11] This is an alarming fact not because entomology and cultural studies are unimportant (they aren't), but because the entire enterprise of civilization depends on humanity stepping back from the ledge. More generally speaking, far more articles and books have been written about the past than the future—about history than futurology. The past is, of course, much easier to study, but the future can be (pro)actively shaped to fit our goals and values. Thus, Robin Hanson asks, "If the future matters more than the past, because we can influence it, why do we have far more historians than futurists?"[12] One hope of mine is that the field of future studies, which explores the *possible*, *probable*, and *preferable* futures of humanity plus *wild cards* (i.e., the "three Ps and a W"), will blossom into a thriving "interdiscipline" in which existential risks in particular become a central focus of urgent philosophical and scientific investigation.[13]

6.2 Development Trajectory

Just for once I'd like to see humans go prepared into one of these giant technological shifts.

—Nate Soares[14]

With respect to the questions just posed—basically, how can humanity avoid an existential catastrophe?—some scholars maintain that preventing specific existential risks is the optimum way to minimize the threat of an existential catastrophe, and thus "that the best giving opportunities must be with charities that primarily focus on reducing existential risk."[15] To quote the former executive director of the Machine Intelligence Research Institute (MIRI), Luke Muehlhauser,

Many humans living today value both current and future people enough that if existential catastrophe is plausible this century, then upon reflection (e.g. after counteracting their unconscious, default scope insensitivity) they would conclude that reducing the risk of existential catastrophe is the most valuable thing they can do—whether through direct work or by donating to support direct work.[16]

In response to those who emphasize "direct work" on existential risks, the philosopher Nick Beckstead has proposed the concept of the world's **development trajectory**. This refers to "a rough summary way the future will unfold over time," where this "summary includes various facts about the world that matter from a macro perspective, such as how rich people are, what technologies are available, how happy people are, how developed our science and culture is along various dimensions, and how well things are going all-things-considered at different points of time."[17] According to Beckstead, direct action to mitigate existential risks is "a promising cause," but it may not be the *most* promising. Rather, there could be "a much broader class of actions which may affect humanity's long-term potential"—that is, a class that includes "very broad, general, and indirect approaches to shaping the far future for the better, rather than thinking about very specific risks and responses."[18]

Beckstead's concept is closely related to the idea of *path dependence* in the social sciences, according to which past events can constrain, in crucial and long-lasting ways, the space of future development. For example, the adoption of the QWERTY keyboard layout in the late 1800s, which was introduced to prevent typing too fast and jamming mechanical typewriters of the time, set technology on a path that would be exceptionally difficult to deviate from today. (Just imagine billions of people having to relearn how to type on a new keyboard layout.) Similarly, there could be decisions that humanity makes today that have long-term consequences for how civilization will evolve in the coming centuries. Perhaps lifting people out of poverty in the developing world will have cascading effects that yield a new generation of philosophers and scientists who devise novel methods for reducing existential risks. Although solving global poverty wouldn't directly reduce existential risks, it could change the configuration of subsequent societies in a way that puts humanity in a better position to survive and thrive.

The same can be said about any number of possible actions: there could be a vast array of *micro-strategies* that are capable of changing our world trajectory in subtle but critical ways. To quote Beckstead at length,

Very persistent trajectory changes that are not existential catas-trophes, could have great significance for shaping the far future. Though it seems unlikely that the far future will inherit many of our institutions exactly as they are, it is not hard to believe that various aspects of the far future—including social norms, values, political systems, and technologies—will be path depen-dent on what happens now, and often in a suboptimal way. In general, it is reasonable to assume that if there is some problem that might exist in the future and we can do something to fix it now, future people would also be able to solve that problem. But if values or social norms change, they might not agree that some things we think are problems really are problems. Or, if a certain standards or conventions get sufficiently entrenched, some problems may be too expensive to be worth fixing.[19]

He adds that

Though thinking about these smaller trajectory changes may be as important as thinking about existential risk, the best ways to address smaller trajectory changes may be very different. For example, it may be reasonable to try to assess, in detail, questions like, "What are the largest specific existential risks?" or, "What are the most effective ways of reducing those spe-cific risks?" In contrast, I would not find it as effective to try to make specific guesses about how we might create smaller posi-tive trajectory changes because there are so many possibilities and many trajectory changes do not have significance that is predictable in advance. No one could have predicted the per-sistent ripple effects that Jesus's life had, for example. In other cases—such as the framing of the U.S. Constitution—it's clear that a decision has trajectory change potential, but it would be hard to specify, in advance, which concrete measures should be taken. Because of this, promising ways to create positive trajec-tory changes in the world may be highly indirect. Improving education, improving our children's moral upbringing, improv-ing science, improving our political system, spreading humani-

tarian values, or otherwise improving our collective wisdom as stewards of the future could create many small, unpredictable positive trajectory changes.[20]

Because the number of ideas, actions, decisions, policies, beliefs, and so on that could nontrivially shape the future is interminable, we will not explore this issue any further. Suffice it to say that people who care about promoting education, morality, science, democracy, humanitarianism, and wisdom could very well have profound positive effects on future civilization. This is a point that is worth underlining, and one that should motivate everyone to do good in all the many domains of life all the time; one never knows exactly how the flapping of a butterfly's wings might alter the weather thousands of miles away.

Instead, our focus in this chapter will be a hodgepodge of **risk mitigation macro-strategies** that more or less *directly* aim to mitigate different existential risks. We can divide these into three general categories: (1) agent-oriented, (2) tool-oriented, and (3) other options. The first includes any mitigation intervention that targets the *agent side* of the agent-tool coupling, whereas the second focuses on the *tool side*. The third is a catch-all category of assorted ideas. While some of the proposals below are quite speculative, I would argue that our existential predicament this century—the 02000s—is sufficiently dire that we should consider a wide range of possibilities, including ones that may initially appear "sci-fi" but could, if they work, have a significant positive impact on our collective adventure into the shadows and mists of things unknown (as I have elsewhere put it).[21] "Thinking big" in the face of monumental challenges should be encouraged rather than dismissed.

So, let's consider (1), (2), and (3) in turn:

6.3 Agent-Oriented Strategies

The two primary strategies here considered are *cognitive enhancement* and *moral bioenhancement*. We will then explore some additional proposals.

6.3.1 Cognitive Enhancement

A **cognitive enhancement** is any process or entity that augments the core capacities of the information-processing machine located between our ears.[22] Cognition consists of multiple subcomponents, such as (a) acquiring information, (b) selecting which information to process, (c) producing mental representations of the world, and (d) retaining information in the form of memories.[23] Thus, a process or entity counts as a cognitive enhancement if and only if it improves the functioning of one or more of these subcomponents.[24] As Nick Bostrom and Anders Sandberg put it, "A cognitively enhanced person . . . is somebody who has benefited from an intervention that improves the performance of some cognitive subsystem without correcting some specific, identifiable pathology or dysfunction of that subsystem."[25]

There are two general versions of cognitive enhancements: *conventional* and *radical*. The former are in widespread use today; examples include caffeine to increase alertness, *ginkgo biloba* and fish oil to improve memory, and mindfulness meditation to enhance concentration, mood, and memory. Another kind of conventional enhancement is education, which essentially provides better mental software to run on the "wetware" of our brains. The result is not just a greater capacity for intellection but changes to the central nervous system itself—e.g., learning to read permanently alters the way the brain processes language.[26]

In contrast, radical cognitive enhancements are those that would produce much more significant changes in cognition, and many are still in the research phase. Consider brain-boosting pharmaceuticals, or **nootropics**. One of the most discussed nootropics in recent years is modafinil, which studies have shown can improve "working memory in healthy test subjects, especially at harder task difficulties and for lower-performing subjects."[27] It is also associated with "significantly enhanced performance on tests of digit span, visual pattern recognition memory, spatial planning, and stop-signal reaction time" (all of which are cognitive tests).[28] This has made it attractive to students trying to earn better grades in school, and indeed one count found that up to 25 percent of students at certain high-ranking universi-

Morality, Foresight, and Human Flourishing

180 • Morality, Foresight, and Human Flourishing

ties have consumed modafinil in an attempt to sharpen their minds.[29] Although there are some drawbacks to this drug, it indicates that appreciable increases beyond normal mental functioning are possible through pharmacological intervention.

Another radical cognitive enhancement is **brain-machine interfaces** (BMIs). This involves connecting an individual's brain to a computer through, for example, a neural implant, thereby enabling her to manipulate external machines or acquire information from the Internet without the intermediary of perception.[30] There is already a great deal of work being done on BMIs, with many notable successes. For example, scientists have trained monkeys to control mechanical arms using *only their minds*, and the first kick at the 2014 World Cup in Brazil was made by a paraplegic named Juliano Pinto who was strapped into a mind-controlled mechanical exoskeleton. Furthermore, Elon Musk has recently hinted that his new company Neuralink is developing a "neural lace" that would allow us to achieve "symbiosis with machines" so that we can "communicate directly with computers without going through a physical interface.[31] But note that accessing information isn't the same as processing it, meaning that someone with a BMI wouldn't necessarily be able to think faster.[32]

There are also genetic interventions that could improve one or more of our core cognitive capacities. For instance, scientists have engineered mice that produce more NR2B subunits of the NMDA receptor, an ion channel protein in nerve cells. As mice age, the NR2B subunits are gradually replaced by NR2A subunits, and this may contribute to the lower neuroplasticity of adult mice brains.[33] After overexpressing the NR2B gene, though, the resulting transgenic mice—nicknamed "Doogie" mice after the fictional prodigy Doogie Howser—were able to "learn faster, remember longer, and outperform [their] wild-type littermates in at least six different behavioral tests."[34] Since mice are "model organisms," this suggests that similar modifications could be made to the NMDA receptors of the human brain.

A final intervention is **iterated embryo selection**. This process involves collecting embryonic stem cells—which are capable of differentiating into every type of cell in the body—from donor embryos.

These cells are then made to differentiate into sperm and ovum (egg) cells. When a sperm and ovum combine during fertilization, the result is a single cell with a full set of genes, called the "zygote." After this occurs, scientists can select the zygotes with the most desirable genomes and discard the rest. The selected zygotes then mature into embryos, from which embryonic stem cells can be extracted and the process repeated. If we understand the genetic basis of intelligence sufficiently well, we could specify selection criteria that optimize for general intelligence. The result would be rapid increases in IQ, a kind of "eugenics" but without the *deeply immoral* consequences of violating people's moral autonomy or introducing more suffering into the world.[35] According to a paper by Nick Bostrom and Carl Shulman, selecting one embryo out of ten, creating ten more out of the one selected, and repeating the process ten times could result in *IQ gains of up to 130 points.*[36] Thus, iterated embryo selection could offer a promising method for creating super-brainy offspring in a relatively short period of time.

Some philosophers argue that cognitive enhancements could have "a wide range of risk-reducing potential," as Bostrom puts it, leading him to argue that "a strong *prima facie* case therefore exists for pursuing these technologies as vigorously as possible."[37] At first glance, this seems right: surely being *smarter* would make us less likely to do something *dumb* like destroy ourselves? But let's take a closer look using the agential risk framework established above.

(i) *How could cognitive enhancements influence agential terror?* Recall that agents in this category are expressly motivated by a malicious intent to harm others. Thus, to mitigate this source of danger, cognitive enhancements would have to interfere with some aspect of the agent's motivations or intentions. And there are reasons for thinking that they could do precisely this.

   Consider apocalyptic terrorists first. Individuals of this sort are inspired by Manichaean belief systems according to which those outside one's religious clique—the infidels, the reprobates, the damned—are perceived as the unholy enemies of all that is good in the universe. This harsh division of humanity into two distinct groups is facilitated in part by a failure to understand "where others are coming from," to see the world from "another person's perspective."[38] Consequently, one could argue that just as literary fiction has expanded many individuals' circle of empathy by educating them about the experiences of others, so too could cognitive enhancements achieve this end by enabling people to gain greater knowledge of other cultures, political persuasions, religious worldviews, and so on.[39] In fact, this line of reasoning leads the bioethicist John Harris to defend the use of cognitive enhancements for moral purposes. As he writes, "I believe that education, both formal and informal, and cognitive enhancement are the most promising means of moral enhancement that are so far foreseeable."[40] The reason is that, he claims, the aversion that some people have toward other belief communities, other races, homosexuals, and so on is not a "brute" reaction in the way that one's fearful reaction to snakes or spiders might be. Rather,

> it is likely to be based on false beliefs about those racial or sexual groups and or an inability to see why it might be a problem to generalize recklessly from particular cases. . . . The most obvious countermeasure to false beliefs and prejudices is a combination of rationality and education, possibly assisted by various other forms of cognitive enhancement.[41]

Thus, by correcting the false beliefs and fallacious inferences of religious extremists, cognitive enhancements could promote more religious moderation, which poses no direct risks to humanity (i.e., religious moderates wouldn't pass the doomsday button test).

Numerous studies have also found a negative correlation between intelligence and religiosity as well as theism.[42] That is to say, people on the high IQ end of the normal distribution curve are less likely to believe in supernatural phenomena or consider themselves religious. This is a mere correlation, though, which could have a hidden common cause, such as average wealth.[43] Yet there are *prima facie* reasons for positing a causal link between aptitude and atheism. For example, one study published in *Sci-*

ence found that analytical thinking (that is, decomposing a problem into its constitutive parts) caused religious disbelief in real-time among subjects in a laboratory. In other words, people who were prodded to think analytically became less likely to describe themselves as religious.[44] Evidence from other domains also suggests that, as one author puts it, "higher levels of intelligence are associated with a greater ability—or perhaps willingness—to question and overturn strongly felt intuitions."[45]

These data suggest that a population of cognitively enhanced individuals would be less religious. This might be desirable because it would presumably yield less religious terrorism, the most dangerous form of terrorism today, and with less religious terrorism one should expect less *apocalyptic terrorism*—a very good outcome from the agential risk point of view.[46] (See also Box 10.)

More generally speaking, Steven Pinker's **escalator hypothesis** states that the observed decline in global violence since the second half of the twentieth century—a trend that subsumes the Long Peace, the New Peace, and the Rights Revolutions—has been driven by rising average IQs in many regions of the world, a phenomenon called the "Flynn effect."[47] For Pinker, the Flynn effect is the crucial catalyst that has "accelerated an escalator of reason and led to greater moral breadth and less violence." The most important concept here is that of "abstract reasoning," which Pinker identifies as being "highly correlated" with IQ. In his words, "The cognitive skill that is most enhanced in the Flynn Effect, abstraction from the concrete particulars of immediate experience, is precisely the skill that must be exercised to take the perspectives of others and expand the circle of moral consideration." He adds that "enhanced powers of reason—specifically, the ability to set aside immediate experience, detach oneself from a parochial vantage point, and frame one's ideas in abstract, universal terms—would lead to better moral commitments."[48] It follows that insofar as cognitive enhancements can extend the intellectual gains of the Flynn effect, they could produce morally superior individuals and therefore a morally superior world.

Box 10. We should register one additional reason that less religious societies might be better. To wish that humanity avoids extinction, one must believe that extinction is possible. But the eschatological narratives of the world's largest religions entail that human extinction is *not* possible. According to these views, some portion of humanity—the righteous, the elect, the true believers—will survive the apocalypse and enter into an eternal paradise with God. There is no scenario in which humans simply cease to be. In fact, an online survey of beliefs about human extinction, conducted by the futures scholar Bruce Tonn, confirms that "Christians and Jews overwhelmingly do not believe that humans will become extinct."* (Muslims were not specifically included in the survey.) It follows that research into and efforts aimed at avoiding human extinction are fundamentally misguided from this perspective. Looking into the future, this could become dangerous because roughly 8 of the 9.3 billion people expected to exist in 2050 will be religious—i.e., a *growing proportion* of humanity may hold views according to which the "naturalistic" survival concerns of existential risk scholars are unworthy of serious attention, or funding.† Consequently, humanity could become even more vulnerable to disasters that, with the right collective efforts, could be effectively neutralized.

* Tonn, Bruce. 2009. Beliefs about Human Extinction. *Futures*. 41(10): 766–773.

† Pew Research Center. 2015. The Future of World Religions: Population Growth Projections, 2010-2050. URL: http://www.pewforum.org/2015/04/02/religious-projections-2010-2050/.

Thus, one might suppose that cognitive enhancements could also mitigate the threat of idiosyncratic actors, many of whom suffer from a marked lack of empathy. If *only* such individuals were more intelligent, if *only* they had higher IQs, perhaps they would be less likely to engage in homicidal acts—which WTDs could soon scale up into omnicidal acts. Indeed, numerous empirical studies have linked cognitive deficits like low IQ and learning disabilities to criminality, meaning that individuals with higher IQs are less inclined to become involved in criminal behavior.[49]

But there is a hitch, since many idiosyncratic actors actually exhibit above-average intelligence and have been fairly well educated. As the psychologist Peter Langman writes about school shooters, in particular, "Contrary to what we might expect, they are not kids who are on the low end of the academic spectrum."[50] For example, Dylan Klebold "spent several years in a program for gifted children" while Eric Harris enjoyed math and science, quoted Shakespeare in his journal, and appears to have read Thomas Hobbes and Friedrich Nietzsche—the latter of whom is widely quoted as saying that "the world is beautiful, but has a disease called Man."[51] Similarly, Adam Lanza was described "as a genius of sorts with a high IQ."[52] And Charles Manson had an IQ of 121, which is considered "highly above average." It follows that augmented IQs alone might not have much of a positive effect on this category of agential risk, meaning that cognitive enhancements might not provide an efficacious solution.[53] Perhaps the most compelling argument that they *could* mitigate idiosyncratic actors is that, despite their intelligence, people like Harris and Manson did hold some notably inaccurate beliefs. For example, one of Harris's motivations was "to kick-start natural selection and eliminate inferior beings."[54] (This is why he wore a shirt that read "Natural Selection" on the day of the massacre.) But this is not exactly how natural selection works! Thus, if he had better understood evolutionary biology, the flaws of social Darwinism, and so on, perhaps he would have been less motivated to carry out his attack.

One encounters similar problems with respect to the two other agential risks not yet discussed, namely, misguided ethicists and ecoterrorists: it is not, or at least not obviously, a lack of psychometric intelligence, abstract reasoning, or veridical beliefs that make these agents risky. For example, one might claim that strong negative utilitarianism (SNU) is false, but on what grounds? SNU consists of two central components: (a) a *consequentialist mandate* to evaluate moral actions based on their consequences, and (b) an *axiological thesis* that specifies the reduction of suffering as the ultimate aim of moral conduct. All forms of utilitarianism accept (a), so let's focus on (b). Now ask: Are there any *facts of the matter* about whether this thesis is correct or not? Would it be the case that if only SNUs could *reason more abstractly* they would recognize (b) as flawed? The answer to both questions appears to be "No."[55] Philosophical arguments for claims like (b) often rely upon thought experiments that characteristically end in, as it were, "the dull thud of conflicting intuitions."[56] When one hears this "dull thud," there is nothing much left to talk about.[57]

The same goes for (some) ecoterrorists. For example, Ted Kaczynski is a Harvard-educated mathematician who wrote about the perils of modern megatechnics eloquently enough to influence people like Bill Joy, the cofounder of Sun Microsystems and author of an influential neo-Luddite manifesto published in *Wired*.[58] However ghastly his crimes were, the Unabomber was not lacking IQ points.[59] Complicating the situation even more is the fact that empirical science unambiguously affirms that the globe is warming and the biosphere wilting due to human activity. Our species *really has* been a monstrously destructive force in the world—perhaps the *most* destructive since cyanobacteria flooded the atmosphere with oxygen some 2.3 billion years ago (see Jennifer Jacquet's quote at the beginning of section 3.2). Thus, the problem with ecoterrorists like the Gaia Liberation Front isn't (generally speaking) that they harbor "false beliefs" about reality. Quite the opposite: their *descriptive world models* are often grounded on solid scientific evidence. Nor are they unable to "set aside immediate experience" or "detach oneself from a parochial vantage point," as Pinker puts it, a topic to which we will return below.

For these reasons, it does not appear that cognitive enhancements would mitigate the agential risks posed by misguided ethicists and ecoterrorists (and possibly idiosyncratic actors). If anything, they could *intensify* such threats by amplifying knowledge about how to kill more people, the ubiquity of human suffering, and anthropogenic environmental destruction.

One final point worth making is that cognitive enhancements would likely increase the rate of technological development, thereby shortening the segment of time between the present and when large numbers of people could have access to a doomsday button. As Persson and Savulescu observe, "The progress of science is in one respect for the worse by making likelier the misuse of ever more effective weapons of mass destruction, and this badness is increased if scientific progress is speeded up by cognitive enhancement."[60] There is also the possibility that cognitive enhancements enable malicious agents to turn themselves into evil geniuses. A fascist authoritarian, for example, might attempt to boost his *instrumental rationality* via cognitive enhancements, thereby empowering him to vanquish his political enemies with even greater facility. An individual so enhanced might even figure out new ways to evade detection (in the case of lone wolves) and/or bring civilization to its knees through violence, subtle manipulation, or blackmail.

So, cognitive enhancements appear to be a "mixed bag" as a person-engineering approach to mitigating agential terror. Given that so little work has been done on this topic, though, one should see the present analysis as the beginning, rather than the end, of the story.

Let's now turn to the flip side of the agential terror-error coin.

(ii) *How could cognitive enhancements influence agential error?* As suggested in subsection 4.3.2, agential error could constitute an even greater threat than agential terror. But it appears that cognitive enhancements could provide a partial solution. For example, consider that higher IQs are positively correlated with a range of desirable outcomes, such as better health, less morbidity, and a lower probability of premature death. One explanation for this correlation is that more intelligent people are less prone to making the sort of cognitive mistakes that can compromise one's health or put one's life in danger. As one study articulates this hypothesis, "Both chronic diseases and accidents incubate, erupt, and do damage largely because of cognitive error, because both require many efforts to prevent what does not yet exist (disease or injury) and to limit damage not yet incurred (disability and death) if one is already ill or injured."[61] In fact, regression analyses show that IQ is a strong *predictor* of fatal automobile accidents, meaning that young men with lower IQs are more likely to die in car crashes than those with higher IQs. Yet another study elaborates the connection as follows:

> *Preventing some aspects of chronic disease is arguably no less cognitive a process than preventing accidents, the fourth leading cause of death in the United States. . . . Preventing both illness and accidents requires anticipating the unexpected and "driving defensively," in a well-informed way, through life. The cognitive demands of preventing illness and accidents are comparable—remain vigilant for hazards and recognize them when present, remove or evade them in a timely manner, contain incidents to prevent or limit damage, and modify behavior and environments to prevent reoccurrence.*[62]

The point is this: if (a) cognitive enhancements increase IQ, and (b) increased IQ is causally linked to better error avoidance, then (c) cognitive enhancements could reduce the threat of agential error. Insofar as this argument is sound, these good effects may be especially pronounced as the world becomes more socially, politically, and technologically complex. For example, another study observes that

> *the advantages conferred by higher levels of g [or IQ] are successively larger in successively more complex jobs, tasks, and settings. Greater experience and other favorable personal traits can compensate to some extent for lower levels of g, but they can never negate the disadvantages of information processing that is slow or error prone.*[63]

Although cognitive enhancements could worsen some types of terror agents, the evidence—albeit indirect—suggests that a population of cognitively enhanced cyborgs would be less susceptible to accidents, mistakes, and errors, and therefore less likely to inadvertently self-destruct in the presence of WTDs.

(iii) *How could cognitive enhancements influence other existential risks not associated with agent-tool couplings?* It seems plausible to say that a smarter overall population would increase humanity's ability to solve a wide range of global problems.[64] Consider Bostrom's calculation that a 1 percent gain in "all-around cognitive performance . . . would hardly be noticeable in a single individual. But if the 10 million scientists in the world all benefited from the drug the inventor would increase the rate of scientific progress by roughly the same amount as adding 100,000 new scientists."[65] Although we noted above that accelerating the pace of science could have disadvantages, it might also put humanity in a better position to neutralize a number of existential risks. For example, superior knowledge about supervolcanoes, infectious diseases, asteroids and comets, climate change, biodiversity loss, particle physics, geoengineering, emerging technologies, and agential risks could lead to improved responses to these threats.[66] In argument form:

(a) Better thinking through the use of cognitive enhancements could improve the quality of scientific research; that is, it could lead to better science.

(b) Better science could yield better theories about the world.

(c) Better theories could enable humanity to more effectively avoid some existential risks.

(d) Thus, cognitive enhancements could enable humanity to more effectively avoid some existential risks.

In the case of radical enhancements that expand our *cognitive space*, we could potentially acquire knowledge about dangerous phenomena in the universe that unenhanced humans could never know about, in principle. In other words, there could be any number of existential risks looming in the cosmic shadows to which we, stuck in our Platonic cave, are "cognitively closed." Perhaps we are in great danger *right now*, but we can only know this if we understand a theory T. The problem is that understanding theory T requires us to grasp a single concept C that falls outside our cognitive space. If an enhancement were to enable one to grasp C, then she or he could potentially devise T and therefore recognize the risk. Only after one recognizes a risk can one invent strategies for avoiding it.

With respect to bad governance, the connection between intelligence and positive outcomes is well-established. As discussed in section 5.2, low integrative complexity in political speeches correlates with a higher probability of war, and integrative complexity is linked to general intelligence. Along these lines exactly, research shows a statistical connection between the IQ of American presidents and how many soldiers die in battle.[67] It follows that political leaders with higher intelligence—perhaps as the result of cognitive enhancements—would be less inclined to cause bloody conflicts with many casualties, and fewer conflicts in the world would reduce the overall probability of a techno-Armageddon.

Finally, one of the most urgent problems in society today is the pervasive lack of basic knowledge about phenomena like climate change and the Anthropocene extinction. Although cognitive capacity is not the same as knowledge, greater cognitive capacity can facilitate the acquisition of knowledge. Thus, the widespread use of cognitive enhancements could foster a more scientifically informed population that could, in turn, be more inclined to take action against these conflict-multiplying context risks. Zooming out, basic knowledge about our evolving existential plight in general is the very first step toward doing something to ensure our continued survival. A civilization whose citizens know every idea in this book, for example, would have a greater chance of attaining technological maturity—or so one might argue. Thus, insofar as cognitive enhancements could produce citizens who understand exactly what the dangers are and what is at stake (e.g., astronomical future value), they could mitigate the cumulative threat of an existential catastrophe.

6.3.2 Moral Bioenhancement

The leading advocates of **moral bioenhancement** are Ingmar Persson and Julian Savulescu, although many other theorists have endorsed the idea. The primary aim of moral bioenhancement is to augment our *motivational urges* to act ethically. Persson and Savulescu argue that this could be achieved by targeting the "core moral dispositions" of (a) *altruism*, and (b) *the sense of justice (or fairness)*.[68] They analyze the former into two components, namely, *empathy* (putting yourself in someone else's shoes) and *sympathetic concern* (caring about the well-being of others). Sympathetic concern in particular is the motivational part, since being motivated to engage in moral behavior requires not just *seeing* the world from the perspective of others but actively *caring* about whether others are happy or sad. For example, when John has sympathetic concern for Jess, it matters to John that Jess is doing okay, and if Jess isn't, John will feel a "pull" to help her. Finally, by the sense of justice, they mean our willingness to engage in reciprocal cooperation with other people, exemplified by the game theoretic "tit-for-tat" strategy of "You scratch my back and I'll scratch yours."[69]

Persson and Savulescu prefer the term "bioenhancement," with the prefix "bio-," because it emphasizes that the interventions under discussion would aim to modify the neural correlates of our moral dispositions through biomedical means. Thus, it is important that the dispositions above are *biologically based*, since if they aren't, biological interventions will be unable to alter them as desired. According to the best current research on animals, identical twins, and the cross-cultural differences between the genders, it appears that both altruism and the sense of justice are indeed biologically based. For example, consider the "ultimatum game," in which a proposer is given a certain amount of money that she or he must divide up with another player, the responder. If the responder accepts the division, then both players are rewarded, but if the responder rejects the offer (e.g., because she or he feels that it is unfair), then neither get any reward. This gives the proposer an incentive to divide the money evenly enough to get the responder's approval while also attending to her selfish desire to allocate more for herself. Thus, as Persson and Savulescu note, researchers

> have found that in the case of identical twins (who share the same genes), there is a striking correlation between the average division with respect to both what they propose and what they

are ready to accept as responders. There is no such correlation in the case of fraternal twins. This indicates that the human sense of fairness has a genetic basis.[70]

They add that "there is also a striking correlation in respect of altruism in identical twins."[71]

So, what sort of biomedical interventions could alter our dispositions in the desired ways? Perhaps some of the same techniques discussed in subsection 6.3.1, including genetic engineering (to produce hyper-moral designer babies) and brain-machine interfaces (that, say, inhibit bad moral impulses). The method most commonly discussed in the literature, though, involves pharmaceuticals. On the model of nootropics, we can call such morality-boosting drugs **mostropics**, since our word "moral" comes from the Latin *mos*, meaning "one's disposition." Although research is still in its infancy, there are hints that drugs could alter our moral characteristics in desirable ways. For example, oxytocin—also known as the "cuddle hormone"—is a naturally occurring hormone and neurotransmitter that is associated with "maternal care, pair bonding, and other pro-social attitudes, like trust, sympathy, and generosity."[72] When administered to test subjects, researchers found that those with elevated oxytocin levels exhibit "significantly more trusting behaviour."[73] There is also evidence that those with more oxytocin in their bloodstream are more trustworthy. Thus, Persson and Savulescu suggest that "in a population with universally elevated oxytocin levels increasing trusting behaviour seems to be matched by increased trustworthiness."[74]

Unfortunately, the benefits of oxytocin appear to be limited to group membership. In one study,

Participants administered oxytocin were significantly more likely to sacrifice a different-race individual in order to save a group of race-unspecified others than they were to sacrifice a same-race individual in the same circumstances. In participants who had been administered a placebo, the likelihood of sacrificing an individual did not significantly depend on the racial group of the individual.[75]

So, oxytocin's morally enhancive effects may be insufficiently generalized to be worth promoting in society today.

The class of drugs known as selective serotonin reuptake inhibitors (SSRIs), which are widely prescribed for depression and anxiety, also exhibit mostropic properties: ingesting them seems "to make subjects more fair-minded and willing to cooperate" in certain situations.[76] For example, one study found that subjects administered the SSRI citalopram (Celexa) were fairer while playing the "dictator game," similar to the ultimate game mentioned above, in which a subject divides money between herself and other participants. Complementing this result is evidence that lower levels of serotonin (as a result of lower levels of its precursor, tryptophan) are correlated with less cooperation in the prisoner's dilemma game.[77]

Although mostropic research is primitive, these studies offer hope that science could someday produce pharmaceuticals with safe and powerful moral effects. But could such drugs help humanity mitigate the threat of an existential catastrophe? Could they enable us to avoid the worst-case scenario of Ultimate Harm? Let's have a look, once again using the agential risk framework.

(i) *How could moral bioenhancements influence agential terror?* The answer to this question appears to depend on whether or not moral bioenhancements are made (a) universally available (perhaps by being injected into the public water supply like fluoride), and (b) compulsory by the state (rather like health insurance is under Obamacare in the United States). The reason is that even a *single* lone wolf or malicious group in the future could acquire sufficient destructive power to unilaterally bring about an existential disaster (call this the condition of *unilateralism*). Unless *everyone* is morally bioenhanced, *no one* will be safe.[78] The obvious problem is that terrorists and sociopaths in particular are among the least likely people to voluntarily use mostropics to alter their moral dispositions.[79] (Sociopaths are, indeed, notoriously resistant to psychiatric treatment.) Thus, even if (a) were the case, a *voluntary regime* of moral bioenhancement would probably not mitigate the threat of agential terror. It would need to be compulsory.

So, what should one expect if conditions (a) and (b) are both satisfied?[80] Here we find some surprising results, as was the case with cognitive enhancements. First, consider apocalyptic terrorism. An integral component of the apocalyptic worldview is a stark dichotomy between Good and Evil—the Manichaean fissure alluded to above that splits humanity into two opposing camps. Frances Flannery refers to this as the condition of "Othering/Concretized Evil." She writes that "this 'Othering' is a conceptual process whereby the 'in-group' (the radical apocalyptic group) ceases to be able to identify in any *empathetic fashion* with 'out-group' members (everyone else)."[81] This same process can lead in-group members to feel little or no sympathy for out-group members: why would one feel motivated to help others if one identifies them as "Concretized Evil"—i.e., the enemies of God who deserve, and will soon justly receive, eternal punishment in hell.[82] Thus, for apocalyptic terrorists, the scope of empathy and sympathetic concern is approximately *coextensive* with their doxastic communities. This is partly what makes the indiscriminate, catastrophic violence unique to religious terrorism appear acceptable from the in-group's perspective: those on the outside are unworthy of eternal life in God's presence, a conclusion reached through a sort of spiritual dehumanization.[83] It follows that, insofar as moral bioenhancements could augment the capacity for empathy and sympathetic concern, they could potentially mitigate the threat of apocalyptic terrorism.[84]

With respect to idiosyncratic actors, the link between moral bioenhancements and this agential risk is relatively straightforward. As the moral philosopher Nicholas Agar writes, "We can imagine a biomedical moral therapy that morally improves a psychopath by restoring a normal aversion to inflicting suffering. Prison psychologists provide moral therapy to psychopaths by talking to them. There's no reason a drug might not have the same moral therapeutic effect."[85] Although Eric Harris was relatively intelligent, he suffered from a sociopathic lack of empathy/sympathy for others. He was, for example, unable to identify with the plights of women, gays, and African Americans, occasionally writing about sending the last group "back to Afrifuckingca were [*sic*] you came from." He also saw himself as different from his peers, once admonishing, "How dare you think that I and you are part of the same species when we are sooooooo different. You aren't human. You are a robot." Elsewhere he wrote that "I feel like God and I wish I was, having everyone being OFFICIALLY lower than me," and "Ich bin Gott," which means "I am God" in German.[86] Thus, insofar as moral bioenhancements could treat psychopathy (or sociopathy), they could have lessened Harris's motivation to kill and perhaps obviated the massacre for which he was responsible. More generally, if moral bioenhancements become widespread in the future, one might expect the prevalence of violent outbursts—e.g., rampage killings and school shootings—to decline, perhaps even disappearing altogether. This could be critical in a world cluttered with WTDs.

Unfortunately, we encounter some problems when it comes to misguided ethicists and ecoterrorists. The reason is that neither appears to suffer from any obvious deficits in their core moral dispositions. For example, SNUs are no less motivated by a "capacity to imagine what it would be like to be another conscious subject and feel its pleasure or pain" or "concern about the well-being of this subject for its own sake" than those who subscribe to alternative ethical systems.[87] Indeed, some SNUs would argue that using a WTD for world-exploder purposes would constitute the *ultimate act of selfless altruism*: after all, one might say, what greater sacrifice is there than killing oneself in the service of eliminating all human suffering in the universe? What greater act is there than destroying every instance of disutility that exists today and could come to exist in the future? If anything, it seems that moral bioenhancements could *exacerbate* this agential risk by increasing the moral motivation of SNUs to destroy the world.[88]

Similarly, one finds altruism and a sense of justice at the heart of many ecoterrorists' ethical beliefs. Consider that individuals of this sort tend to embrace a circle of empathy/sympathy that extends far beyond the human species to include many, most, or all other living organisms—even the "Gaian system" as a whole.[89] This is in part what underlies the biospheric egalitarianism of deep ecology, according to which "all living things are alike in having value in their own right."[90] They also tend to see the damage caused by human overpopulation, overexploitation, habitat destruction, pollution, climate change, and so on, as a specifically *moral* catastrophe, given the suffering that these phenomena have inflicted on other creatures. In the most radical cases, some ecoterrorists argue that the total extermination of *Homo sapiens*, the primary culprits of climate change and the Anthropocene extinction, would constitute the supreme manifestation of justice in Persson and Savulescu's sense: we have destroyed the environment, so now we too must be destroyed.[91] It follows that, once again, moral bioenhancements could potentially *worsen* this type of agential risk by reinforcing the ecoterrorist's conviction that the living Earth must be saved from "Homo shiticus."

Thus, our conclusion is that moral bioenhancements are also a "mixed bag." Agents who suffer from a lack of altruism or the sense of justice would clearly benefit from such interventions, but there are other categories of risky agents whose behaviors stem from strong moral convictions based on, it seems, precisely these dispositions. Another point is that, since apocalyptic terrorists and idiosyncratic actors would be unlikely to use moral bioenhancements voluntarily yet misguided ethicists and ecoterrorists might actually seek them out, *either* moral bioenhancements should become universal and compulsory *or* they should not be made available at all.[92] The worst situation would be one in which moral bioenhancements are widely available but not compulsory, as this would fail to mitigate the risks of apocalyptic terrorists and idiosyncratic actors while aggravating the risks of misguided ethicists and ecoterrorists.

(ii) *How could moral bioenhancements influence agential error?* The answer to this hinges upon the extent to which it has any moral dimension, given that agential error stems from unintended accidents, mishaps, gaffes, and blunders. In other words, we can put the question like this: would a perfectly moral person be less likely to make a mistake?

Perhaps there is a sense in which caring more about people in distant countries (across space) or future generations (across time) could lead one to take extra steps to avoid an error. For example, someone might say, "Because I deeply value future people, I'm going to be *especially careful* so that no accidents occur." This could result in situations that are less vulnerable to catastrophic mistakes: e.g., one might install additional safeguards to prevent an unintended nuclear launch, laboratory leak, or nanotech spill, this step being explicitly motivated by a sense of altruism and justice. On a global scale, the effects of moral bioenhancement on agential error could add up. But this appears to be the only sense in which such enhancements could diminish the relevant threat.

(iii) *How could moral bioenhancements influence other existential risks not associated with agent-tool couplings?* A central thesis defended by Persson and Savulescu is that while cognitive enhancements are an important component of moral improvement, they are not *sufficient* for agents to behave ethically. That is to say, it is not enough to merely *know* what the right action is; one must also be *inspired* to pursue that action. As Persson and Savulescu put it, referring to Pinker's escalator theory,

> We do not want to deny that enhanced powers of reason are tremendously important for moral enhancement—perhaps Pinker is right that they are the main force behind the moral improvements that he lists. But we do want to deny that once reason "is programmed with a basic self-interest and an ability to communicate with others, its own logic will impel it, in the fullness of time, to respect the interests of ever-increasing numbers of others."[93]

The reason they deny Pinker's claim is because they "do not see how such an expanding circle of concern is possible without the assistance of the moral dispositions of altruism and a sense of justice." After all, the combination of reason and self-interest could very well lead one to engage in treacherous acts, such as robbing and killing "an injured stranger in the wilderness rather than help[ing] him" or abstaining "from returning a favour to someone you will not ever see again rather than to return it at a cost to yourself."[94] Without sympathetic concern to motivate one's behaviors, they argue, no one would help anyone else unless it were expedient for the helper.

The relevance of these points is that in part (iii) of subsection 6.3.1, we explored some reasons for thinking that cognitive enhancements could reduce the probability of an existential catastrophe associated with natural phenomena, bad governance, and large-scale human action. But perhaps knowing that a phenomenon P carries a risk is insufficient to make one *do something* about P. If Persson and Savulescu are correct, cognitive enhancements won't save humanity from the Ultimate Harm of an existential catastrophe unless they are accompanied by the enhancement of our core moral dispositions.[95] In fact, the primary global risk that they believe a cocktail of moral and cognitive enhancers could mitigate is climate change. The reason is that augmenting the scope of our altruism toward spatially and temporally distant humans, making people more trustworthy and trusting of others, and improving our willingness to engage in reciprocal cooperation (specifically, to reduce the carbon footprint of industrial civilization) could enable humanity to extricate itself from the game theoretic death trap of the commons tragedy. In their words,

> we think that sympathy and a sense of justice are indispensable for being fully moral, and that the explanation of why humanity so far has failed to deal with climate change and environmental destruction—in spite of the enhanced powers of reason—is that they leave self-interest untouched and call upon our insufficient sympathy and sense of justice as regards future generations and non-human animals.[96]

The only way, it seems, for moral bioenhancements to effectively mitigate climate change is if their use is universal and compulsory, since otherwise society would face the problem of free riding.[97] Yet this would potentially magnify the threats posed by certain types of agential risks, as previously discussed. Future research should thus explore the net benefits, all things considered, of these various options.

6.3.3 Other Options

There are a few additional agent-oriented strategies that we should consider before moving on to the next section. Some of these have already been implied in previous discussion, but they are worth making explicit here:

(a) *Mitigate the environmental triggers that contribute to terrorism.* We noted in subsection 4.3.3 that radical change breeds radical religion—a proposition that Ackerman suggests could be generalized to "radical change breeds radical beliefs."[98] Since global warming will very likely cause radical change, we should expect the frequency and size of apocalyptic terrorist groups to increase in the future.[99] Along these lines, if environmental destruction becomes more salient this century, it could (and should) fuel more intense concern for the biosphere—just consider the emotional effects of widely shared pictures of emaciated polar bears and plastic debris in the stomachs of birds—and this could, in turn, elevate the background threat posed by ecoterrorists (unfortunately). Thus, I would argue that abating environmental triggers—i.e., mitigating context risks—should constitute an *exigent top priority* for institutions and governments around the contemporary world. The national security problems of tomorrow are being incubated by climate apathy and denialism today.

(b) *Improve social conditions.* Many idiosyncratic actors emerge from life situations associated with family problems, social isolation, recent personal losses, and so on. In a world replete with WTDs, it could become critical to minimize such conditions: a failure to do so need only produce a single individual with access to a doomsday button for the great experiment of civilization to suddenly end. Perhaps this means implementing better "social safety nets" to keep people from reaching "rock-bottom"—i.e., societies should move toward more "democratic socialist" systems, as exemplified by some European countries—or establishing more robust involuntary mental health services for psychotic individuals who "pose a serious risk of physical harm to themselves or others."[100]

(c) *Use mass surveillance to track dangerous agents.* This is a controversial idea advocated by Persson and Savulescu in lieu of coercing or tricking terrorists into using moral bioenhancements. As they write,

> *To counteract the threat of highly destructive attacks from such groups, liberal democracies have to avail themselves of the sophisticated means of surveillance that modern technology offers. Such surveillance will make these democracies less liberal, but . . . the xenophobia that results from a terrorist attack from some ethnic group is an even greater threat to liberalism.*[101]

The glaring drawback of surveillance is that it could be abused by autocratic leaders. As Bostrom notes, "Improved governance techniques, such as ubiquitous surveillance . . . , might cement [a dangerous] regime's hold on power to the extent of making its overthrow impossible."[102] One possible response to this is what the scholars Steve Mann, Jason Nolan, and Barry Wellman call "sousveillance." This involves the citizens themselves monitoring agents of the state through the use of wearable cameras and other recording apparatuses. Thus, the surveillees (those being watched) surveil the surveillers (those doing the watching), a kind of "inverse panopticon" that could help protect individuals from overreach and misconduct by the government.[103] If such a system were implemented, one might argue that society could get to "have its cake and eat it to": surveillance by the state could reduce the threat of lone wolves, while sousveillance by the citizenry would ensure that those in charge don't abuse their power. Indeed, at the extreme this could foster a completely "transparent society," as David Brin calls it, in which *everyone can see what everyone else is doing*. However creepy this may seem, it would almost certainly be preferable to one in which only Big Brother is watching our every move.[104]

(d) *Solve the AI control problem.* That is to say, figure out how to create a friendly superintelligence whose value system aligns with ours, or is evolutionarily constrained by some "meta-value" that conduces to human well-being. While an unfriendly superintelligence would, for reasons already established, probably guarantee annihilation, a friendly superintelligence could offer something like an *existential panacea*—or what the scholars Owen Cotton-Barratt and Toby Ord call a "eucatastrophe."[105] To paraphrase Stephen Hawking, if superintelligence isn't the worst thing to happen to our species, then *it will probably be the best.*[106]

Imagine, for example, a civilization guided by the foresight of a superintelligent mind; imagine that this superintelligence were to establish a singleton—that is, a global governing system—whose policies are shaped by its super-human wisdom? Not only could this singleton neutralize nearly all the threats posed by nature, but it could prevent violent conflicts from breaking out and solve the cooperation problems driving climate change and biodiversity loss. With respect to agential risks, if humans are small children playing with flamethrowers (as suggested in section 4.1) and if we are unable to "grow up" through the use of cognitive or moral enhancements, then a superintelligence could act as our parent, making sure that we don't burn down the global village either on purpose or on accident. Thus, a friendly superintelligence—being benevolent by definition—could implement a mass surveillance system to keep a watchful eye on bad individuals or groups. Additionally, it could invent new, safe, and effective human enhancements that, among other things, make our species less of a threat to itself. As for error, Bostrom writes that

> superintelligence would also eliminate or reduce many anthropogenic risks. In particular, it would reduce risks of accidental destruction, including risk of accidents related

to new technologies. Being generally more capable than humans, a superintelligence would [also] be less likely to make mistakes, and more likely to recognize when precautions are needed, and to implement precautions competently. A well-constructed superintelligence might sometimes take a risk, but only when doing so is wise.[107]

But perhaps we shouldn't be too roseate about this possibility. After all, it was our ingenuity—our *genius*—as big-brained, tool-using primates that made possible the invention of artifacts that now threaten our own survival. Thus, it could be that a superintelligence creates novel technologies that introduce brand new risks into the world, and that these risks make the superintelligence vulnerable to an existential catastrophe of its own. There may, indeed, be some types of problems that require a level 10 of intelligence to create but a level 11 of intelligence to solve. A fruit fly, for example, is intelligent enough to find its way into a half-empty wine bottle but often fails to figure out a way to escape this death trap. Perhaps a superintelligence will be so smart that it invents the ultimate doomsday technology that, once extracted, cannot be put back into the urn (see section 5.3).

(e) *Stop shouting into the sky.* Messaging to extraterrestrial intelligence (METI) is sort of like parachuting into the jungle during the Vietnam War and, unsure about whether the enemy is around, shouting "Over here!" The less electromagnetic radiation that humanity sends into the cosmos, whether on purpose or not, the better off we will be.

* * *

Finally, one of the important reasons for creating a taxonomy of agential risks is that different types of agents may be more or less likely to couple themselves with different types of technologies. For example, a "misguided ethicist" (as we have termed this phenomenon) who believes that the ultimate aim of moral action is to eliminate human suffering would find a WTD with a very high probability of *total destruc-*

tion in one fell swoop far more attractive than a WTD that, if something minor goes wrong or circumstances aren't exactly right, could yield an *endurable* catastrophe of some sort that does the opposite of eliminating suffering, instead *amplifying* it. For example, consider the many unlikely contingencies that must hold for a designer pathogen to kill *every human* on Earth or the thousands of nuclear weapons that would need to be detonated to induce a nuclear winter—and even then there would be no guarantee of total annihilation. In contrast, far fewer conditions need to hold for an exponentially growing swarm of self-replicating nanobots to spread around the globe and destroy the biosphere. Thus, weaponized nanotech may offer a more reliable, and therefore more attractive, way of causing extinction than either an engineered pandemic or a nuclear attack. It follows that an omnicidal SNU should prefer this option over the other two.

Such analyses, which have received little scholarly attention, are important because they could help agential risk experts fructify their efforts to mitigate a catastrophe. In the case above, attention should focus more on misguided ethicists acquiring advanced molecular nanotechnology than building a few nuclear weapons, since the grey goo scenario appears to have "all-or-nothing" consequences, and SNU entails that if you can't eliminate everyone, you shouldn't eliminate anyone. Similar analyses could deliver essential insights about which WTDs apocalyptic terrorists, idiosyncratic actors, ecoterrorists, and machine superintelligences might prefer, given the cognitive and moral properties unique to each.[108]

6.4 Tool-Oriented Strategies

Another option is to target the tool side of the agent-tool coupling, since, as mentioned above, the capacity of agents to destroy civilization is limited by the means at their disposal. Much of the mitigation efforts by institutions and governments since World War II has focused on preventing weapons from getting into the hands of "evildoers," to quote George W. Bush, whether such actors are state or nonstate. For example, the Treaty on the Non-Proliferation of Nuclear Weapons (i.e., the Non-Proliferation Treaty), "aims to prevent the spread of nuclear weapons and weapons technology, to foster the peaceful uses of nuclear energy, and to further the goal of disarmament."[109] There is also the 1925 Geneva Protocol, which bans the use of chemical and biological weapons. This was later expanded by the Biological Weapons Convention of 1972, which prohibits the development and stockpiling of biological weapons, as well as the Chemical Weapons Convention of 1993, which prohibits the development and stockpiling of chemical weapons. Moving forward, the international community may benefit from similar treaties concerning the use of nanoweapons and artificial intelligence, both of which could be much more dangerous than chemical or nuclear weapons.[110]

Along these lines, some scholars have suggested that humanity should pursue the **broad relinquishment** of certain dangerous technologies.[111] This is, to some extent, what the treaties above attempt to do, and with some degree of success, although there are reasons for worrying about the poor current state of the Non-Proliferation Treaty. As the philosopher Mark Walker points out, relinquishing certain technologies may ultimately prove otiose in part because of how arduous—nay, impossible—it is to contain *information*, which, as the maverick Stewart Brand once said, "wants to be free."[112] Focusing on the biological sciences in particular, Walker explains that

> *relinquishment requires us to not only stop future developments but also to turn back the hands of time, technologically speaking. If we want to keep ourselves completely immune from the potential negative effects of genetic engineering we would have to destroy all the tools and knowledge of genetic engineering. It is hard to imagine how this might be done. For example, it would seem to demand dismantling all genetics labs across the globe and burning books that contain information about genetic engineering. Even this would not be enough since knowledge of genetic engineering is in the minds of many.[113]*

There is also a concern that imposing moratoriums on whole fields of inquiry would force dual-use research underground rather than stopping it, and research conducted under such conditions would be even more hazardous than if it were to occur in properly regulated spaces.[114] Consequently, Walker concludes that he "would rate the chances for relinquishment as a strategy pretty close to zero."[115]

But an inability to relinquish entire domains of research doesn't mean that we can't manage future technological developments *to some extent*. Even if the macroscopic trends of technology are beyond human control—just as the movement of a flock of starlings is beyond the control of any single bird, a central idea of Langdon Winner's *autonomous technology thesis*[116]—there may still be lower-level phenomena that we can effectively manipulate. This suggests a strategy that Ray Kurzweil calls **fine-grained relinquishment**.[117] As he writes,

> *I do think that relinquishment at the right level needs to be part of our ethical response to the dangers of twenty-first-century technologies. One constructive example of this is the ethical guideline proposed by the Foresight Institute: namely, that nanotechnologists agree to relinquish the development of physical entities that can self-replicate in a natural environment.[118]*

A related idea involves trying to influence the *order of arrival* of different technologies. This could be accomplished by varying how much money is allocated to different fields, where a deficit of funds would result in slower progress and an abundance would produce the opposite. The order of arrival is important because it could reduce the threat of dangerous technologies without imposing categorical restrictions on them, which is what relinquishment entails. For example, imagine that the technologies X and Y will both introduce novel existential risks, but technology Y could mitigate the risks of technology X whereas the reverse is not the case. Which technology should we prefer to confront first? Obviously, technology Y, since this scenario would yield a lower overall level of danger. This is the concept of **differential technological development**. As Bostrom writes, "What matters is not only *whether* a technology is developed, but also *when* it is developed, by *whom*, and in *what context*." Thus, humanity should attempt to "retard the development of dangerous and harmful technologies, especially ones that raise the level of existential risk;

and accelerate the development of beneficial technologies, especially those that reduce the existential risks posed by nature or by other technologies."[119]

Consider the concrete case of molecular nanotechnology (technology X) and superintelligence (technology Y). Both pose novel dangers to human survival. But their relation is asymmetrical: if we create a friendly superintelligence, it could help neutralize the risks posed by molecular nanotechnology, whereas molecular nanotechnology probably won't help us create a friendly superintelligence—indeed, it could cause the premature arrival of superintelligence by enabling us to manufacture extremely powerful supercomputers, thereby giving AI researchers less time to solve the control problem.[120]

The point is that there could be top-down interventions into certain fields of science that strive to change not the ultimate outcome of which technologies are developed, but the chronology according to which these technologies become a reality.

A final tool-oriented strategy involves the development of **defensive technologies**. For example, to combat the threat of self-replicating nanobots (grey goo), Kurzweil argues that "we will ultimately need to provide a nanotechnology-based planetary immune system"—i.e., "nanobots embedded in the natural environment to protect against rogue self-replicating nanobots." Let us call such nanobots "blue goo." According to Kurzweil, "This immune system would have to be capable of contending not just with obvious destruction but with any potentially dangerous (stealthy) replication, even at very low concentrations," and it would "ultimately require self-replication; otherwise it would be unable to defend us."[121]

In response to this idea, Joy and others have raised the concern that a self-replicating blue goo immune system could itself pose a grave threat due to the possibility of an "autoimmune disorder," whereby the defensive nanobots malfunction and destroy the "organism"—our planet—that they were supposed to be protecting. But Kurzweil counters that this doesn't mean that we should refrain from building a planetary immune system. In his words, "No one would argue that humans would be better off without an immune system because of the potential of developing autoimmune diseases. Although the immune system can itself present a danger, humans would not last more than a few weeks (barring extraordinary efforts at isolation) without one."[122]

The possibility of defensive technologies also provides another reason to oppose the aforementioned option of broad relinquishment. As the nanotech gurus Robert Freitas and Ralph Merkle write,

> *Attempts to block or "relinquish" molecular nanotechnology research will make the world a more, not less, dangerous place. This paradoxical conclusion is founded on two premises. First, attempts to block the research will fail. Second, such attempts will preferentially block or slow the development of defensive measures by responsible groups. One of the clear conclusions reached by [one author] was that effective countermeasures against self-replicating systems should be feasible, but will require significant effort to develop and deploy. . . . But blocking the development of defensive systems would simply insure that offensive systems, once deployed, would achieve their intended objective in the absence of effective countermeasures.*[123]

Consequently, Freitas and Merkle endorse a version of differential technological development: "Actively encouraging rapid development of defensive systems by responsible groups while simultaneously slowing or hindering development and deployment by less responsible groups ('nations of concern')," they write, "would seem to be a more attractive strategy [than] blocking the development of defensive systems."[124] So, there are a number of options here that could lessen the dangers associated with advanced technologies, although further research is desperately needed.

6.5 Other Strategies

The proposals below do not specifically target agent-tool couplings, nor is this list intended to be exhaustive.

(i) *Space colonization*. The geographical dispersal of a population is positively correlated with its survival—that is, the more spread out a species is, the lower its probability of extinction. This rule applies, it seems, no less to the three-dimensional realm of outer space than to the two-dimensional (curved) world of our oblate spheroid, third rock from the sun. Thus, Jason Matheny argues that "colonizing space sooner, rather than later, could reduce extinction risk."[125] This echoes a "common sense" belief held by many leading intellectuals. For example, Stephen Hawking states that he doesn't "think the human race will survive the next thousand years, unless we spread into space," but that "once we spread out into space and establish independent colonies, our future should be safe."[126] The former NASA administrator Michael Griffin similarly claims that "human expansion into the solar system is, in the end, fundamentally about the survival of the species."[127] And Elon Musk, the founder of SpaceX, asserts that "there is a strong humanitarian argument for making life multi-planetary . . . in order to safeguard the existence of humanity in the event that something catastrophic were to happen."[128] Finally, Derek Parfit writes that

> *What now matters most is how we respond to various risks to the survival of humanity. We are creating some of these risks, and discovering how we could respond to these and other risks. If we reduce these risks, and humanity survives . . . , our descendants or successors could end these risks by spreading through this galaxy.*[129]

For perhaps a majority of those who contemplate humanity's long-term future, space colonization offers the strongest reason for believing that the bottleneck hypothesis may be true. By expanding to Mars and beyond, catastrophes that are sequestered to a single planet, like Earth, won't threaten the perpetuation of our posthuman lineage. (Only a few scenarios could affect multiple planets at once, such as the vacuum bubble disaster or an unfriendly superintelligence that launches spacecraft into the universe to destroy all other civilizations.) Colonizing space could also be relatively cheap, offering arguably the best "bang for the buck" compared to most other macro-strategies on the marketplace of ideas. And launching humans to Mars, for example, would involve a lot of scientific knowledge that is already well-established, unlike, say, developing reliable mostropics, which awaits one or more major breakthroughs in neuropharmacology and related fields. In fact, there are good reasons for expecting our species to colonize space by the middle of this century. For instance, NASA operates a colonization program that hopes to put humans on Mars by the 2030s, and Musk announced in 2016 that SpaceX would establish a Martian colony "in our lifetimes."[130] The firmament is the last great frontier, and it appears to be only a matter of decades before we make it our abode.

But one should not be too Panglossian about this option. To paraphrase the political scientist Daniel Deudney, space colonization could potentially initiate a cascade of undesirable outcomes that result in major catastrophic or existential threats to humanity.[131] For example, expanding into space would almost certainly entail its militarization—a process that first began in the mid-1940s, when Germany designed a ballistic missile that traveled beyond Earth's atmosphere. This could increase the probability of, as Deudney puts it, "interworld/interspecies" wars, perhaps involving the use of large asteroids as ammunition (see below). Furthermore, colonization of our and other galaxies would make regulating dangerous technologies far more onerous. If, through a process of "adaptive radiation," new species evolve that are hostile to our lineage, they could accumulate large arsenals of WTDs for *otherworld*-destroying purposes. Making matters worse, the *very threat* of this could, for game theoretic reasons, lead otherwise peaceable civilizations to destroy the sovereign planetary nations around them, and beyond, in order to avoid being destroyed themselves. Deudney also notes that securing Earth from hostile aliens might require a "planetary garrison state" that could elevate the likelihood of an oppressive totalitarian state taking control of our planet.

So, there are a number of serious problems with this proposal. As the theoretical physicist Freeman Dyson writes, "When mankind moves out from earth into space, we [will] carry our problems with us."[132] It could be that spreading beyond Earth— the "planetary cradle of civilization"—could actually render our existential predicament *more* precarious.

(ii) *Track near-Earth objects.* This is a "no-brainer," as Neil deGrasse Tyson suggests below. Asteroids and comets constitute one of the few existential risks that advanced technologies could easily reduce or eliminate. If astronomers were to identify an assailant from the heavens speeding our direction, they could launch a spacecraft to deflect it away from Earth—although such a spacecraft could also be exploited to direct an impactor *toward* the planet, a problem known as the "deflection dilemma."[133] Another possibility would be to nuke the asteroid or comet into fragments, although this has the disadvantage of raining smaller pieces over regions of the planet that may be populated.

Given how easy it would be to protect ourselves against asteroids and comets, Tyson writes that "if humans one day become extinct from a catastrophic collision, we would be the laughing stock of aliens in the galaxy for having a large brain and a space program, yet [meeting] the same fate as [those] pea-brained, space program-less dinosaurs that came before us."[134] While extinction-causing impactors are rare, they should not be ignored.

(iii) *Geoengineering.* While stratospheric geoengineering poses a number of risks, it could also protect humanity against the catastrophic effects of sudden climate change. On the one hand, as noted earlier, geoengineering with sulfate aerosols would be quite cheap. One estimate reports that "the annual cost of stratospheric aerosols could be less than $10 billion per year, which is orders of magnitude less than the costs of climate change mitigation strategies."[135] Thus, this strategy could be pecuniarily feasible for a wide range of actors (both state and nonstate), which could be good news if civilization were to find itself in a "climate emergency" that demands immediate action.[136] Indeed, as the Global Challenges Foundation argues, if a last-minute maneuver like this were successful, then the *failure* to use geoengineering could itself "constitute a global catastrophic risk."[137]

To be sure, any meddling with the highly complex, chaotic climatic system should be avoided if at all possible. But given current trends in ongoing carbon emissions output as well as the rise to power of political groups (such as the Republicans in the United States) that reject the established conclusions of climatological science, geoengineering ought not be too quickly dismissed as a macro-strategic option for obviating a calamity. Indeed, as a recent article about the first real-world atmospheric geoengineering experiment (set to occur over the Arizona desert by early 2018) notes, "as risky as geoengineering may seem at the moment, many scientists appear to be cautiously accepting the idea that we should study the underlying science in more detail."[138]

(iv) *Bunkers, or refuges.* The economist Robin Hanson argues that subterranean bunkers—also called "refuges"—could offer a promising way to survive certain catastrophic scenarios. Such bunkers could house a group of people for extended periods, or they could be occupied at all times, just as the Svalbard Global Seed Vault, also known as the "Doomsday Vault," remains constantly stocked with seeds. Hanson suggests that bunkers should include things like "libraries, machines, seeds, and much more," but he notes that certain types of capital might be worthless after a major disruption. For example, cell phones, computers, and medical instruments require specific infrastructure to function properly, and such infrastructure would likely be destroyed in the event of a global disaster. Thus, a better choice of items would be those that could enable bunker occupants to return to a Paleolithic hunter-gatherer lifestyle in the post-catastrophe world (in hopes of later developing agricultural and industrial economies). This leads Hanson to suggest that "it might make sense to stock a refuge with real hunter-gatherers and subsistence farmers, together with the tools they find useful."[139]

The bunker itself would need to withstand not only the initial catastrophe but any subsequent attacks from individuals who happened to survive. In Hanson's words, "If desperate people try-ing to survive a social collapse could threaten a refuge's long-term viability, such as by looting the refuge's resources, then refuges might need to be isolated, well-defended, or secret enough to sur-vive such threats." This could be achieved by establishing "secret rooms deep in a mine, well stocked with supplies, with some way to monitor the surface and block entry."[140]

In addition to isolation and secrecy, as Seth Baum, David Denkenberger, and Jacob Haqq-Misra write, a successful bunker would need to be self-sufficient, desirable, pleasant, accessible (if a bunker isn't easily accessible, then it should have a continuous population), equipped with the material resources needed to re-build civilization, and occupied by a sufficient founder popula-tion.[141] The last is important because a group that survives a ca-tastrophe in a bunker would be unable to repopulate the planet if they were to lack a certain degree of genetic diversity.[142] According to the ecologist Philip Stephens, "You need 50 breeding individu-als to avoid inbreeding depression and 500 in order to adapt"—al-though the futurist Karim Jebari argues that "to achieve a colony that could survive a global catastrophe and eventually repopulate earth, we would need *at least* 80 colonists."[143] As for the number of people needed to sustain a closed bunker over long periods of time, the anthropologist John Moore writes in a NASA study that 160 people in total are required for space missions lasting up to 200 years.[144]

In addition to subterranean bunkers, some scholars have explored the possibility of *aquatic* and *extraterrestrial* refuges. Unfortunately, as Baum and his colleagues note, "Aquatic refuge design has received virtually no attention," although submarines could provide a template for what such a refuge might look like.[145] With respect to extraterrestrial refuges, one can distinguish be-tween two basic types, namely, *planetary* and *autonomous*. The former includes extraterrestrial colonies, which could range in size from large metropolises to towns of a few hundred. The lat-ter would be spacecraft that are completely surface-independent of any astronomical body. Such bunkers could contain the absolute minimum amount of technology necessary for their "rescue agents" to live comfortable lives, since less technological complexity generally entails a lower probability of malfunction. They would also have the advantage of protecting their inhabitants from events that could destroy subterranean, aquatic, and planetary bunkers, such as the grey goo and strangelet disaster scenarios (see sections 4.2.2 and 3.3, respectively).

Admittedly, the idea of building bunkers to avoid an existential catastrophe does not exactly inspire enthusiasm—it lacks "sex appeal," unlike, say, space colonization and cyborgization. But as Baum, Denkenberger, and Haqq-Misra argue, "Refuges could . . . be the difference between the long-term success or failure of human civilization on Earth and beyond. For this reason, refuges merit consideration within the broader landscape of possible responses to catastrophic threats to humanity."[146] Yet other risk scholars are less sanguine about this idea, arguing that bunkers would, all things considered, offer little extra benefit relative to a range of GCR scenarios. (See Figure I.) As Beckstead writes about subterranean and aquatic bunkers in particular,

> Refuges may initially seem like a reliable method of ensuring that civilization will recover from a wide range of potential catastrophes. However, on closer inspection, many existing systems serve similar functions [e.g., government-funded bunkers for private citizens, "continuity of government" bunkers, private bunkers, and submarines], and refuges would have limited impact for many potential catastrophes. Many proposed catastrophes render refuges of limited use [for] simple reasons: for "overkill" catastrophes, people in refuges cannot survive anyway (these include alien invasions, runaway AI, global ecophagy from nanotechnology, physics disasters like the "strangelet" scenario, and simulation shutdowns); "underkill" catastrophes probably are not destructive enough for refuges to be relevant (these include earthquakes and hurricanes); and refuges are largely irrelevant to long-term environmental dam-

Figure I. Possible Outcomes after a Catastrophe

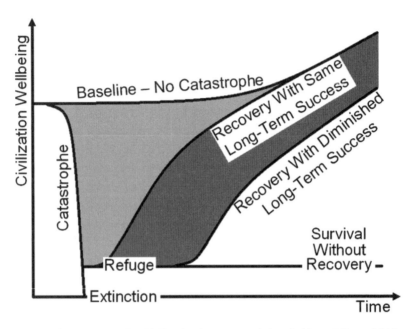

Source: Seth Baum, David Denkenberger, and Jacob Haqq-Misra. 2015. Isolated Refuges for Surviving Global Catastrophes. *Futures*. 74: 45–56.

> *age scenarios (these include climate change, gamma-ray bursts, supernovae).*[147]

Although Beckstead's argument is rather persuasive, more research is needed on this topic—especially on the possibility of autonomous extraterrestrial bunkers, since these could sidestep at least some of the objections that Beckstead propounds.

(v) *A preemptive strike.* In section 1.4, we likened the present moment to a narrow foundation that supports an infinitely tall skyscraper (with all future humans living in it). If current humans fail to prevent an existential catastrophe, it would be tantamount to the foundation imploding, thus causing the entire edifice to come crashing to the ground. Now, imagine that you see someone chipping away at the foundation, or placing explosives next to it. Given the astronomical value thesis and related considerations about future generations, you decide that drastic action is necessary. After all, you tell yourself, from an ethical perspective, the good done by saving the skyscraper would *far exceed* the bad done by murdering an omnicidal nutcase who wants to unilaterally foreclose humanity's vast future prospects. So you "dirty your hands," as some moral and political theorists would say, by aiming your rifle and shooting—a truly horrible act that you find unbearably distressing, but one that can, at least arguably, be justified on moral grounds.[148] This is the basic line of reasoning behind the use of extreme measures—such as a preemptive nuclear strike—in those rare situations where we (a) are sufficiently certain that an actor is about to cause an existential catastrophe, and (b) feel sufficiently confident that extreme measures would save the world.[149]

In considering this option, Bostrom writes that

a preemptive strike on a sovereign nation is not a move to be taken lightly, but in [exceptionally dangerous cases]—where a failure to act would with high probability lead to existential catastrophe—it is a responsibility that must not be abrogated. Whatever moral prohibition there normally is against violating national sovereignty is overridden in this case by the necessity to prevent the destruction of humankind. Even if the nation in question has not yet initiated open violence, the mere decision to go forward with development of the hazardous technology in the absence of sufficient regulation must be interpreted as an act of aggression, for it puts the rest of . . . the world at an even greater risk than would, say, firing off several nuclear missiles in random directions.[150]

The neuroscientist Sam Harris also suggests that there are particular circumstances in which the preemptive use of nuclear weapons could be morally justified. Focusing on the threat posed by violent Islamists, he argues,

> *What will we do if an Islamist regime, which grows dewy-eyed at the mere mention of paradise, ever acquires long-range nuclear weaponry? If history is any guide, we will not be sure about where the offending warheads are or what their state of readiness is, and so we will be unable to rely on targeted, conventional weapons to destroy them. In such a situation, the only thing likely to ensure our survival may be a nuclear first strike of our own. Needless to say, this would be an unthinkable crime—as it would kill tens of millions of innocent civilians in a single day—but it may be the only course of action available to us, given what Islamists believe.[151]*

One cannot overemphasize that *this option must remain a last resort.* Even a single nuclear weapon exploding somewhere around the world would indelibly alter the course of human history, and probably not for the best. Nonetheless, as Bostrom argues, it is equally "crucial that we make room in our moral and strategic thinking for this contingency," given what would be lost if an existential catastrophe were to occur.[152]

* * *

Lastly, there could be some *combination* of macro-strategies (as well as development-trajectory-changing micro-strategies) that produce net positive results. Perhaps the use of advanced cognitive and moral enhancements by our spacefaring descendants could obviate the militarization concerns that Deudney articulates. Or, maybe differential technological development—along with what Muehlhauser calls "differential intellectual progress," whereby "our progress on the philosophical, scientific, and technological problems of AI safety outpace our progress on the problems of AI capability such that we develop safe superhuman AIs before we develop arbitrary superhuman AIs"—could lead not only to the arrival of superintelligence before molecular nanotechnology, but to the successful creation of a human-friendly superintelligence, which then governs humanity through a sort of "benevolent global hegemony."[153] Various combinatorial configurations of strategies implemented in parallel could have additive or even synergistic effects—effects that yield optimal arrangements for squeezing through the bottleneck of heightened hazards before us. This is a topic of paramount importance that, as indicated above, requires further (and immediate) research.

Chapter 7: Concluding Thoughts

7.1 Doom Soon?

Without the possibility of a future, there is nothing left but despair.
Thus, if we give up on the future, we give up on ourselves.
—Wendell Bell[1]

Let us recap the most general observations and theses of the book so far. They are: The number of existential risk scenarios has increased significantly since the middle of last century, and the Doomsday Clock has steadily inched closer to midnight—phenomena that are, at the very least, *consistent with* humanity approaching a Great Filter. Furthermore, there are reasons for expecting certain agential risks to become more dangerous in the coming decades, and the window for meaningful action on climate change and the Anthropocene extinction is rapidly sliding shut.[2] Making matters worse, the macro-strategies explored in the previous chapter do not offer the sort of reassurance that one should hope for: many have notable downsides while others, such as aquatic bunkers, are unlikely to be taken seriously by politicians. We have no track record of surviving the threats before us, and the Great Silence is a constant and noisy reminder that, at least ostensibly, *life almost never makes it beyond our current state of technological development.*

Taken together, these data suggest that our prior probability estimates of an existential catastrophe ought to be disturbingly high. As Bostrom expounds in his original 2002 article on existential risks,

The balance of evidence is such that it would appear unreasonable not to assign a substantial probability to the hypothesis that an existential disaster will do us in. My subjective opinion is that setting this probability lower than 25% would be misguided, and the best estimate may be considerably higher.[3]

This speculation aligns with the other estimates of section 1.1, which, readers may recall, place the probability of an existential disaster occurring this century between 19 and 50 percent.[4] It also comports with Stephen Hawking's recent remarks that we live in the most perilous period of human history, ever. In his words,

Now, more than at any time in our history, our species needs to work together. We face awesome environmental challenges: climate change, food production, overpopulation, the decimation of other species, epidemic disease, acidification of the oceans. Together, they are a reminder that we are at the most dangerous moment in the development of humanity. We now have the technology to destroy the planet on which we live, but have not yet developed the ability to escape it.[5]

Some philosophers, though, believe that human extinction is *even more likely* than empirical analyses indicate. Enter the **doomsday argument**, which has been most vigorously defended and elaborated by John Leslie.[6] This attempts to show, through *a priori* reasoning, that we are systematically underestimating the probability of annihilation. To be clear, the conclusion is *not* that doom is imminent but rather that we should inflate our prior estimates of doom, whatever they happen to be. The idea can be understood through a simple analogy. Imagine two buckets filled with balls. Bucket A has 10 balls numbered 1 through 10, whereas bucket B has 1,000 balls numbered 1 through 1,000. Your job is to pick a random ball from one of the buckets and then guess which bucket it came from, A or B.[7] So you reach in and grab a ball with the number 8. Given that your chances of pulling out an 8 from bucket A are significantly higher than from bucket B (which contains far more possibilities), you guess that the ball is from bucket A—and you are almost certainly right.

With this example in mind, consider two hypotheses about how many humans will ever come to exist in the universe. Hypothesis A says that this number is 100 billion, whereas hypothesis B specifies it as 100 trillion. Now, picture *yourself* as a randomly selected person from all the people who will ever be born. We know as a matter of fact that your number in this series is about 60 billion, since this is roughly the number of people who have previously lived on Earth—in other words, if you were a ball pulled from a bucket, your number would say "60 billion." So, given this information, which hypothesis should you favor, A or B? The best answer appears to be A, despite it being considerably less optimistic than B. This is the essence of the doomsday argument: annihilation is more probable than one would otherwise think.

The crucial premise of this argument is that "one should reason as if one were a random sample from the set of all observers in one's reference class," where the relevant reference class here is the total population of humans across both space *and time*. This is called the "self-sampling assumption" (SSA), and, to be sure, not all philosophers accept it.[8] Some prefer the "self-indication assumption" (SIA), which states that "given the fact that you exist, you should (other things [being] equal) favor hypotheses according to which many observers exist over hypotheses on which few observers exist."[9] In other words, if there are two possible worlds, X and Y, where X contains 10 billion people and Y contains 100 trillion, you should reason as if you are in Y rather than X. While SIA essentially makes the Doomsday Argument "go away," it carries a doomsday implication of its own.[10] The reason is this: if we should reason as if we exist in a world that includes many observers like us, then we should expect the Great Filter to be *in front of* rather than *behind* us, since a Great Filter in our past would imply few observers at our stage of development.[11] It follows that, as Robin Hanson puts it, those who accept SIA should "drastically increase" their estimates of an existential catastrophe.[12]

The assumptions of SSA and SIA, one could argue, are the best ideas that philosophers have devised about how to generate self-locating beliefs.[13] Yet both suggest that Ultimate Harm lurks in our future, perhaps not too far away. This should give us *extra pause* when think-ing about what dangers may dot the road ahead.[14]

7.2 Two Types of Optimism

> *Being on the right side of history is no less important*
> *than being on the right side of futurology.*

Finally, when one takes seriously the many ideas of this book, both scientific and philosophical, it is easy to fall prey to counterproduc-tive reactions like *panic*, at one extreme, and *defeatism*, at the oth-er—where either can sometimes lead to the other. In a short article for the *Edge*, Jennifer Jacquet coins the term "anthropocebo effect" to refer to "a psychological condition that exacerbates human-induced damage—a certain pessimism that makes us accept human destruc-tion as inevitable."[15] This can arise in situations in which people feel overwhelmed by a problem, such as climate change, for which hu-manity is causally responsible. But rather than respond with proac-tive eagerness to find a solution, one becomes bogged down by the dismal implications. The anthropocebo effect is a very real hindrance that attends the study of existential risks: it is all too easy to throw one's hands up and declare the conundrums too formidable and, con-sequently, our situation hopeless. One must suppress this impulse whenever it emerges, because however dire the situation may appear, the future—like the melody of a song unfinished—has not yet been written.

It would be helpful here to draw a distinction between what the economist Paul Romer describes as "two very different types of opti-mism."[16] First, *complacent optimism* is passive in nature: it anticipates that things will work themselves out somehow, so there's no need to worry. Second, *conditional optimism* is active in nature: it recognizes a deficiency and then motivates one to fix it. Thus, given these defini-tions, one can be intellectually pessimistic about the future and still embody a sort of invigorating hope for the future. (As the novelist Cormac McCarthy quips, "I'm a pessimist but that's no reason to be gloomy!"[17]) In Romer's words,

Pessimism is more likely to foster denial, procrastination, apathy, anger, and recrimination. It is conditional optimism that brings out the best in us. So we should stop saying that "the end is near." We should say instead: "Ok, we made some mistakes. We can start fixing them by pointing our innovative efforts in a slightly different direction. If we do, we can do things that are even more amazing than the truly amazing things we have already accomplished. It will be so easy that looking back it will seem painless. Let's get going."[18]

The purpose of existential risk studies is to map out the obstacle course of natural and anthropogenic hazards before us and, following our sagacious guide of intelligent anxiety (section 1.7), figure out how best to slalom around these threats. The invisible hand of time inexorably pushes us forward, but the direction in which we move is not entirely outside our control.[19]

* * *

Have we fallen into a mesmerized state that makes us accept as inevitable that which is inferior or detrimental, as though having lost the will or the vision to demand that which is good?

—Rachel Carson, *Silent Spring*

Postscript

Nearly all of this book was written over the course of three and a half intensive weeks following the 2016 U.S. presidential election (although the book was subsequently edited and revised for three additional months; as the aphorism goes, "all writing is rewriting"). It was motivated by an acute sense that humanity stands at a critical crossroads in its career and that a failure to take seriously the warnings of "scientific eschatology," if one will, could have truly disastrous consequences for perhaps the only species in the universe with our level of intelligence. This book attempts to lay out something like a "paradigm" for understanding and neutralizing the swarm of risks buzzing before us. I have no idea if it has succeeded.

But I should also like to say something about my own considered opinions on these topics, which were largely hidden from view for the sake of objectivity. For reasons discussed in Box 2, namely, the *last few people problem*, I suspect that human extinction within the next few centuries is relatively improbable, although the danger is big enough to warrant serious concern. Just consider the uncontacted tribes in the Brazilian Amazon, the 80 to 150 people stationed in Villa Las Estrellas, Antarctica, and the small towns in northern Siberia. It would take a truly *planetary* catastrophe to eliminate every last terrestrial astronaut on spaceship Earth. This being said, I also think that the probability of a global disaster, perhaps one resulting in an existential catastrophe (as defined in section 1.2), is distressingly high—*at least* 25 percent within the next 100 years but probably closer to, or exceeding, Martin Rees's 50-50 estimate. There are several trends that

225

push me, most reluctantly, toward this unwanted conclusion. One is that we increasingly live in a world marked by the coexistence of both *archaic worldviews* and *neoteric artifacts*. That is to say, it appears that ancient normative beliefs about what reality is like and, more importantly, *how it ought to be* are on a collision course with advanced technologies that possess the destructive power to permanently abort civilization. This is an incredibly dangerous situation. I also worry a great deal about the *ostensibly* less minatory danger of agential error: perhaps "posthuman error" will pose less of a threat, but in the meantime we have to worry about *human error* causing an "oopsy daisy" existential mishap. And lastly, I do not consider myself an optimist with respect to superintelligence: (a) I recognize, I think clearly, that there are *far* more ways to get the control problem wrong than right (see subsection 4.3.1), and (b) it seems to me that superintelligence is inevitable, at some point, barring a major societal disruption between now and then. In sum, I see many of the hallmarks that one should expect to see if a Great Filter were located in our future. I suspect, maybe, that other civilizations have been in exactly our situation before but didn't live to tell their tale.

More than anything, I hope this book has contributed to the cultivation of a safer threat environment. I would be elated if my relatives in 02100, or beyond, were to glance through this manuscript and say, "Well, none of *that* happened—not even close!" This is the best-case scenario; the second-best is that they add, "Although things could have turned out very differently if scholarly tomes like this, sounding the alarm in a levelheaded way, grounded by a 'robust sense of reality,' had never been published."[1] In service of this end, I also hope this book will encourage what I would call *rhizomatic scholarship*, especially on futurological issues. By this I mean scholarship that preferentially favors *breadth* over *depth*, insofar as limited resources (like time and memory) entail an inverse relation between the two. As the technology maverick Kevin Kelly writes in a 2008 blog post, information is the "fastest growing entity" in the universe today, an observation that is consistent with the academy becoming more specialized and human knowledge more fragmented.[2] The result is an exponential rise in what one could call "relative human ignorance," which is a measure of the disparity between what any given individual knows (or could possibly know) and what the collective whole knows (e.g., as recorded in academic textbooks).[3] In this sense, people are more ignorant today than ever before in human history; as I have written elsewhere, *everyone today knows almost nothing about most things*.[4]

This is a problematic predicament because it means that there are lots of people rowing the boat but no one steering—and it appears that we might be headed toward an outcrop of jagged rocks. Now more than ever, I believe we need brave intellectuals willing to embrace a kind of *bold dilettantism* (or, if one prefers a less disparaging term, *bold polymathy*) that strives to see the forest more than the trees. Students of philosophy—my own chosen field—may be particularly well-suited for this task; indeed, some metaphilosophical theories characterize philosophers as "reflective generalists" who are tasked with, as the pragmatist Wilfrid Sellars famously wrote, "understand[ing] how things in the broadest possible sense of the term hang together in the broadest possible sense of the term."[5] I tried to make this book, insofar as my limited abilities allowed, an inspirational exemplar of bold dilettantism—that is, of rhizomatic scholarship—given the tremendous diversity of ideas that it brings together. Moving forward, to fully grasp our situation on the planet, in the universe, and at this strange moment in history, we need *many more* scholars scanning the scraggly topography of collective human knowledge. We need *many more* scholars interested in trading a bit of depth—which has its costs, I can affirm—for breadth. While contemporary academia tends to frown upon such a methodological stance, this only means that young people have a golden opportunity to do what students do best with respect to reigning paradigms: rebel.

At the very least, I implore readers to take with them the simple maxim that opened this book: "Be curious and care!" This is, indeed, derived from my own normative perspective on the world, which can mostly be reduced to two distinct things: curiosity and kindness. The first compels one to never stop questioning, interrogating, doubting, and investigating the views of others, as well as one's own precious beliefs, while the second ensures that such criticism always remains constructive rather than destructive, positive rather than negative.

There is nothing better to be than a curious and kind person. Perhaps we shall someday achieve a society in which the actions of all people are guided by these attitudes. Here's to hope—hope without fear, at the risk of platitude.

Acknowledgments

Thanks to Gary Ackerman, Nicholas Agar, Nick Beckstead, Shankar Bhamidi, David Brin, Ryan Carey, George Church, Frances Flannery, Robin Hanson, Manu Herrán, Jennifer Jacquet, Daniel Kokotajlo, Sean Mannion, David Pearce, Ingmar Persson, Martin Rees, Susan Schneider, Peter Singer, Andrew Snyder-Beattie, Bruce Tonn, Steven Umbrello, and Roman Yampolskiy for comments on the book. Any remaining errors in the text are mine and mine alone. Thanks especially to my partner Dr. Whitney Trettien; she is the most lovely, wonderful, funny, incredible, inspiring, entertaining, intelligent, hilarious, insightful, comforting, creative, beautiful, and brilliant person that I have ever known.

Notes

Preface

1   See Sagan, Carl. 1983. Nuclear War and Climate Catastrophe: Some Policy Implications. *Foreign Affairs*. 62(2): 257–292.

2   Note that other estimates are far, *far* higher. For example, Nick Bostrom estimates that some 10^{16} people with normal lifespans could exist on Earth before it becomes uninhabitable (due to the sun), assuming a relatively stable global population close to current numbers. But there could be an astronomically greater number if future people were to colonize space. See section 1.4, as well as Bostrom, Nick. 2013. Existential Risk Prevention as Global Priority. *Global Policy*. 4(1): 15–31; and Bostrom, Nick. 2014. *Superintelligence: Paths, Dangers, Strategies*. Oxford: Oxford University Press, Box 7.

3   See Baum, Seth, and Anthony Barrett. 2015. The Most Extreme Risks: Global Catastrophes. In Vicki Bier (editor), *The Gower Handbook of Extreme Risk*. Farnham, U.K.: Gower. Baum and Barrett add that "Sagan's 500 trillion number may even be an underestimate."

4   Kuhn, Thomas. 1962. *The Structure of Scientific Revolutions*. Chicago, IL: University of Chicago Press.

5   See, e.g., Sam Harris's TED talk on superintelligence, which helped introduced the topic to a large audience outside academia: Harris, Sam. 2016. Can We Build AI without Losing Control over It? TED. URL: https://www.ted.com/talks/sam_harris_can_we_build_ai_without_losing_control_over_it.

6   Although there are some excellent books in the general vicinity, such as Häggström, Olle. 2016. *Here Be Dragons: Science, Technology and the Future of Humanity*. Oxford: Oxford University Press.

7   As a point of reference, this book is written at about the same level as the

average *Stanford Encyclopedia of Philosophy* article, whose target demographic is advanced undergraduates and graduate students.

8   There is also an oceanic wealth of information in the endnotes, which readers are therefore encouraged to peruse. Indeed, many topics that have yet to be explored, analyzed, dissected, and elaborated by existential risk scholars are identified as such in the endnotes—i.e., calling all young scholars!

9   The field of future studies is sometimes said to be about "the three Ps and a W"—that is, about the possible, probable, and preferable futures, plus wild cards. This is a topic discussed later on in the book.

Chapter 1

1   Pinker, Steven. 2011. *The Better Angels of Our Nature: Why Violence Has Declined*. New York, NY: Penguin Books. See also Shermer, Michael. 2015. *The Moral Arc: How Science Makes Us Better People*. New York, NY: Henry Holt, about which Pinker says, "If you wanted a sequel to *The Better Angels of Our Nature*, one which explored all of our spheres of moral progress, not just the decline of violence, this is it."

2   A particularly insightful comment about moral progress across time comes from an interview I conducted with Pinker, who writes that, "As for collective moral progress, I see it as pushed and pulled by two sides of human nature. Dragging us back are atavistic mindsets like zero-sum thinking, authoritarianism, tribalism, dominance, and vengeance, which operate pretty much by default. Pulling us forward are the better angels of our nature like empathy, self-control, and reason, which are energized by the Enlightenment institutions of democracy, science, education, open economies, and a global community." See Torres, Phil. 2016. The United States Is Not an Apocalyptic Wasteland, Explains Steven Pinker. *Motherboard*. URL: https://motherboard.vice.com/en_us/article/steven-pinker-talks-about-donald-trumps-victory-long-term-progress-and-wheth.

3   The cosmologist Stephen Hawking emphasizes the *unique seriousness* of our existential situation in a 2016 article for the *Guardian*. See section 7.1 for Hawking's thesis.

4   As Andrew Snyder-Beattie writes in this very context, "The evidence [for a decline in violence] is compelling, but Pinker's story is one of aggregates, not of individuals. And in a world where individuals could kill millions, even one person outside the reach of an empathetic upbringing or the Leviathan is too many. To keep humanity's trajectory peace-

ful, it will need driven and empowered individuals on the other side as well—those with the ambition to make humanity's future as bright as possible." See Snyder-Beattie, Andrew. 2015. Small Groups, Dangerous Technology: Can They Be Controlled? *Bulletin of the Atomic Scientists.* URL: http://thebulletin.org/small-groups-dangerous-technology-can-they-be-controlled8270.

5   A theorist inclined to project historical trends into the future might hypothesize something like an "existential risk singularity" in the coming centuries. Along Kurzweilian lines, this would refer to a moment beyond which the introduction of new existential risk scenarios (from exponential technology) becomes so rapid that humanity can no longer keep up with, make sense of, or defend against this new, phantasmagoric threat environment. See subsection 4.2.1 for related issues.

6   For an overview of rigorous methods for estimating the probability of human extinction, see Tonn, Bruce, and Dorian Stiefel. 2013. Evaluating Methods for Estimating Existential Risks. *Risk Analysis.* 33(10): 1772–1787.

7   Leslie, John. 1996. *The End of the World: The Science and Ethics of Human Extinction.* New York, NY: Routledge.

8   Sandberg, Anders, and Nick Bostrom. 2008. Global Catastrophic Risks Survey. Technical Report. URL: http://www.global-catastrophic-risks.com/docs/2008-1.pdf.

9   Rees, Martin. 2003. *Our Final Hour: A Scientist's Warning. How Terror, Error, and Environmental Disaster Threaten Humankind's Future in This Century—on Earth and Beyond.* New York, NY: Basic Books. Note that other reputable scholars have proposed similar figures. For example, the microbiologist Frank Fenner, whose virological work helped to eliminate smallpox, predicted in 2010 that "humans will probably be extinct within 100 years, because of overpopulation, environmental destruction and climate change." The legal scholar Richard Posner describes the risk of extinction this century as "significant." The Canadian biologist Neil Dawe says that "he wouldn't be surprised if the generation after him witnessed the extinction of humanity." The ecologist Guy McPherson claims that humanity will follow the dodo into the evolutionary grave by 2026. The geophysicist Brad Werner famously asked during a talk "Is Earth Fucked?," to which he answered "more or less." And, as we will discuss more in section 7.1, Nick Bostrom puts the probability of an existential catastrophe at 25 percent *or higher*. See Edwards, Lin. 2010. Humans will be extinct in 100 years says eminent scientist. PhysOrg.com. URL: https://phys.org/news/2010-06-humans-extinct-years-

eminent-scientist.html, and Jamail, Dahr. 2013. The Coming "Instant Planetary Emergency." *Mother Jones*. URL: http://www.motherjones. com/environment/2013/12/climate-scientist-environment-apocalypse-human-extinction.

10   Insurance Information Institute. Mortality Risk. Accessed on December 12, 2016. URL: http://www.iii.org/fact-statistic/mortality-risk.

11   I discuss these comparisons in several previous publications, such as Torres, Phil. 2016. How Likely Is an Existential Catastrophe? *Bulletin of the Atomic Scientists*. URL: http://thebulletin.org/how-likely-existential-catastrophe9866; and Torres, Phil. 2016. Existential Risks Are More Likely to Kill You Than Terrorism. Future of Life Institute. URL: https://futureoflife.org/2016/06/29/existential-risks-likely-kill-terrorism/. Note also that Bostrom writes the following: "It is possible that a publication bias is responsible for the alarming picture presented by these opinions. Scholars who believe that the threats to human survival are severe might be more likely to write books on the topic, making the threat of extinction seem greater than it really is. Nevertheless, it is noteworthy that there seems to be a consensus among those researchers who have seriously looked into the matter that there is a serious risk that humanity's journey will come to a premature end." See Bostrom, Nick. 2009. The Future of Humanity. Jan-Kyrre Berg Olsen, Evan Selinger, and Soren Riis (editors), *New Waves in Philosophy of Technology*. New York, NY: Palgrave Macmillan.

12   According to a National Consortium for the Study of Terrorism and Responses to Terrorism (START) report, an average of 9 Americans were killed in a terrorist attack each year between 2004 and 2013. This contrasts with a NOAA report finding that an average of 31 people die from lightning strikes each year. Obviously, both numbers are less than the 91 deaths per year average from the National Research Council for asteroid collisions with Earth. See START. American Deaths in Terrorist Attacks. Accessed on March 23, 2017. URL: https://www.start.umd.edu/pubs/START_AmericanTerrorismDeaths_FactSheet_Oct2015.pdf; NOAA. How Dangerous is Lightning? Accessed on March 23, 2017. URL: http://origin-www.nws.noaa.gov/om/lightning/odds.shtml. See also Torres, Phil. 2016. Existential Risks Are More Likely to Kill You Than Terrorism. Future of Life Institute. URL: https://futureoflife.org/2016/06/29/existential-risks-likely-kill-terrorism/.

13   For a helpful explanation, see Spiegelhalter, David. 2012. How Likely Am I to Be Hit by an Asteroid? BBC. URL: http://www.bbc.com/future/story/20120222-waiting-for-a-rock-to-fall.

14 See also Auerbach, David. 2015. A Child Born Today May Live to See Humanity's End, Unless... *Reuters*. URL: http://blogs.reuters.com/great-debate/2015/06/18/a-child-born-today-may-live-to-see-humanitys-end-unless/.

15 See *Bulletin of the Atomic Scientists*. Timeline. Accessed on December 12, 2016. URL: http://thebulletin.org/timeline.

16 See *Bulletin of the Atomic Scientists*. 2016. It Is Still Three Minutes to Midnight. URL: http://thebulletin.org/it-still-three-minutes-midnight9107.

17 *Bulletin of the Atomic Scientists*. 2017. It Is Two and a Half Minutes to Midnight. URL: http://thebulletin.org/sites/default/files/Final%20 2017%20Clock%20Statement.pdf.

18 Krauss, Lawrence, and David Titley. 2017. Thanks to Trump, the Doomsday Clock Advances toward Midnight. *New York Times*. URL: https://www.nytimes.com/2017/01/26/opinion/thanks-to-trump-the-doomsday-clock-advances-toward-midnight.html?_r=0.

19 One possible exception is Bruce Tonn, who writes, referring to himself, "It is the opinion of the author that the probability of human extinction is probably fairly low, maybe one chance in tens of millions to tens of billions, given humans' abilities to adapt and survive." Note that Tonn is discussing not to an existential catastrophe *in general*, but human extinction *in particular* (one of several types of existential risk). As explored in the postscript, one could estimate a low probability of extinction in the next few centuries while simultaneously maintaining that an existential catastrophe is highly probable. See Tonn, Bruce. 2009. Beliefs about Human Extinction. *Futures*. 41(10): 766–773.

20 See Bostrom, Nick. 2002. Existential Risks: Analyzing Human Extinction Scenarios and Related Hazards. *Journal of Evolution and Technology*. 9(1).

21 The "if we wish" qualification corresponds to a moral value, which yields a proposed civil right, called "morphological freedom." As Bostrom puts it, "Transhumanists promote the view that human enhancement technologies should be made widely available, and that *individuals should have broad discretion over which of these technologies to apply to themselves* (morphological freedom)." Bostrom, Nick. 2005. In Defense of Posthuman Dignity. *Bioethics*. 19(3): 202-214. Italics added.

22 See Bostrom, Nick. 2008. Why I Want to be a Posthuman When I Grow Up. In Bert Gordijn and Ruth Chadwick (editors), *Medical Enhancement and Posthumanity*. Berlin: Springer; and Bostrom, Nick. 2003.

Transhumanist Values. *Review of Contemporary Philosophy*. 4. URL: http://www.nickbostrom.com/ethics/values.html.

23   Note that *one does not need to be a transhumanist* to care about existential risks, nor does one need to be a transhumanist to care about global catastrophes. Indeed, we should expect all morally normal people to consider the topic of "global catastrophic risk reduction" to be extremely important, given the huge amounts of human suffering such an event would cause. The present book adopts the transhumanist framework because (a) much of the work on existential risks has been done by transhumanists, and (b) the present author believes that, barring a major disruption, the descriptive component of transhumanism is more or less inevitable, and that as person-engineering technologies become increasingly common in society, their acceptance will rise as well, as has so often occurred with past technologies. (See, e.g., Johnson, Brian David. 2012. The Four Stages of Introducing New Technologies. *Slate*. URL: http://www.slate.com/articles/technology/future_tense/2012/01/new_technologies_enter_our_lives_and_society_in_four_stages_.html.) See also endnote 34.

24   Precautionary principle. Dictionary.com. Accessed on February 9, 2017. URL: http://www.dictionary.com/browse/precautionary-principle. An important related idea comes from Cass Sunstein, who proposes what he calls the "Catastrophic Harm Precautionary Principle" to take the place of the precautionary principle, which he argues is incoherent. This alternative principle states that "when risks have catastrophic worst-case scenarios, it makes sense to pay special attention to those risks, even when existing information does not enable regulators to make a reliable judgment about the probability that the worst-case scenarios will occur." Sunstein describes a "more aggressive" form of his principle as follows: "Regulators should consider the expected value of catastrophic risks, even when it is highly unlikely that those risks will come to fruition. In assessing expected value, regulators should consider the distinctive features of catastrophic harm, including the 'social amplification' of such harm. Regulators should choose cost-effective measures to reduce those risks and should attempt to maximize net benefits. Regulators should also create a 'margin of safety' for catastrophic risks, also with cost-effective measures; the extent of the margin of safety should [be] chosen with reference to its anticipated benefits and costs." See Sunstein, Cass. 2007. The Catastrophic Harm Precautionary Principle. *Issues in Legal Scholarship*. 6(3). See also Sunstein, Cass. 2009. *Worst-Case Scenarios*. Cambridge, MA: Harvard University Press.

25   More, Max. 2004. The Proactionary Principle, Version 1.0. URL: http://

www.extropy.org/proactionaryprinciple.htm. One might also note here what I call the "paradox of human enhancement," which states: The very intelligence and wisdom we seek through cognitive and moral enhancements is precisely what is needed in order to know whether tampering with human biology in such a way is ultimately for the best. This paradox, it seems, casts a faint shadow over the proactionary principle.

26 Bostrom, Nick. 2012. Existential Risk Prevention as Global Priority. *Global Policy*. 4(1): 15–31. In contrast, a bioconservative might stipulate the definition of an existential risk as "one that either (a) threatens the extinction of *Homo sapiens*—that is, through total annihilation or the creation of a replacement posthuman population—or (b) causes a permanent and drastic reduction in of our quality of life." Yet another definition comes from the FHI scholars Owen Cotton-Barratt and Toby Ord, according to which "an existential catastrophe is an event which causes the loss of a large fraction of expected value." Cotton-Barratt, Owen, and Toby Ord. 2015. Existential Risk and Existential Hope: Definitions. Future of Humanity Institute Technical Report. URL: http://www.fhi.ox.ac.uk/Existential-risk-and-existential-hope.pdf. Finally, it is worth noting that not all existential risk scenarios are equally bad, nor does the concept of an existential risk exclude some future situations in which there exists an *astronomical amount of suffering*. As the researchers David Althaus and Lukas Gloor write, Bostrom's definition of existential risks "is unfortunate in that it lumps together events that lead to vast amounts of suffering and events that lead to the extinction (or failure to reach potential) of humanity. However, many value systems would agree that extinction is not the worst possible outcome, and that avoiding large quantities of suffering is of utmost moral importance. We should differentiate between existential risks (i.e., risks of 'mere' extinction or failed potential) and risks of astronomical suffering ('suffering risks' or 's-risks'). S-risks are events that would bring about suffering on an astronomical scale, vastly exceeding all suffering that has existed on Earth so far. The above distinctions are all the more important because the term 'existential risk' has often been used interchangeably with 'risks of extinction,' omitting any reference to the future's quality. Finally, some futures may contain both vast amounts of happiness *and* vast amounts of suffering, which constitutes an s-risk but not necessarily a (severe) x-risk. For instance, an event leading to a future containing 10^{35} happy individuals and 10^{25} unhappy ones, would constitute an s-risk, but *not* an 'x-risk.'"(Some paragraph breaks were removed.) Althaus, David, and Lukas Gloor. 2016. Reducing Risks of Astronomical Suffering: A Neglected Priority. Foundational Research Institute. URL: https://foundational-research.org/reducing-risks-of-astronomical-suf-

fering-a-neglected-priority/. My own view is that the s-risk concept is an important insight, and thus deserves serious consideration alongside x-risks.

27   Ibid. Elsewhere, Bostrom writes that a posthuman civilization is one that exhibits at least one of the following: (a) population greater than 1 trillion persons, (b) life expectancy greater than 500 years, (c) large fraction of the population has cognitive capacities more than two standard deviations above the current human maximum, (d) near-complete control over the sensory input, for the majority of people for most of the time, (e) human psychological suffering becoming rare occurrence, (f) any change of magnitude or profundity comparable to that of one of the above. Quoted from Bostrom, Nick. 2009. The Future of Humanity. In Jan-Kyrre Berg Olsen, Evan Selinger, and Soren Riis (editors), *New Waves in Philosophy of Technology*. New York, NY: Palgrave Macmillan.

28   These are presented in Bostrom, Nick. 2002. Existential Risks: Analyzing Human Extinction Scenarios and Related Hazards. *Journal of Evolution and Technology*. 9(1); Bostrom, Nick, and Milan Ćirković. 2008. Introduction. In Nick Bostrom and Milan Ćirković (editors), *Global Catastrophic Risks*. Oxford: Oxford University Press; and Bostrom, Nick. 2012. Existential Risk Prevention as Global Priority. *Global Policy*. 4(1): 15–31.

29   See Bostrom, Nick, and Milan Ćirković. 2008. Introduction. In Nick Bostrom and Milan Ćirković (editors), *Global Catastrophic Risks*. Oxford: Oxford University Press.

30   Indeed, I have elsewhere argued that it is both *conceptually incoherent*, since the y-axis mixes spatial and temporal categories (a "category error"), and *empirically inadequate*, since it is unable to accommodate risks like germline mutations and aging.

31   I have previously discussed these issues in Torres, Phil. 2015. Problems with Defining Existential Risk. *Ethical Technology*. URL: http://ieet.org/index.php/IEET/more/torres20150121; Torres, Phil. 2015. A New Typology of Risks. *Ethical Technology*. URL: http://ieet.org/index.php/IEET/more/torres20150205; and Torres, Phil. 2016. *The End: What Science and Religion Tell Us about the Apocalypse*. Charlottesville, VA: Pitchstone Publishing. Also, notice that the difference between the generational and transgenerational categories is the difference between talking about people *within* the global population and talking about the global population *itself*.

32   See Persson, Ingmar, and Julian Savulescu. 2012. *Unfit for the Future: The Need for Moral Enhancement*. Oxford: Oxford University Press. Al-

though perhaps there are some worst-case scenarios that don't clearly fall under the "existential risk" umbrella, such as arbitrary eternal damnation of all humanity (see the end of section 5.3). Thanks to Daniel Kokotajlo for pointing this out.

33 See Bostrom, Nick. 2002. Existential Risks: Analyzing Human Extinction Scenarios and Related Hazards. *Journal of Evolution and Technology*. 9(1).

34 A helpful concept here is that of a *premortem analysis*. This involves imagining that some not-yet-executed plan was executed and failed, and then trying figure out why. As the economist Richard Thaler writes, "Before a major decision is taken, say to launch a new line of business, write a book, or form a new alliance, those familiar with the details of the proposal are given an assignment. Assume we are at some time in the future when the plan has been implemented, and the outcome was a disaster. Write a brief history of that disaster." See Thaler, Richard. 2017. The Premortem. Edge.org. URL: https://www.edge.org/response-detail/27174.

35 Taleb, Nassim. 2005. *The Black Swan: Why Don't We Learn that We Don't Learn?* New York, NY: Random House.

36 See Bostrom, Nick. 2012. Existential Risk Prevention as Global Priority. *Global Policy*. 4(1): 15–31.

37 See Bostrom, Nick, and Milan Ćirković. 2008. Introduction. In Nick Bostrom and Milan Ćirković (editors), *Global Catastrophic Risks*. Oxford: Oxford University Press.

38 See Global Challenges Foundation. 2016. Global Catastrophic Risks 2016. URL: http://globalprioritiesproject.org/wp-content/uploads/2016/04/Global-Catastrophic-Risk-Annual-Report-2016-FINAL.pdf.

39 Whatever "sufficiently severe" means in this context—a topic for further discussion.

40 That is, not including Leslie's estimate of a 30 percent chance of doom within the next 500 years. Note also that some risks, such as those associated with superintelligence, appear to be all-or-nothing (i.e., if a superintelligence kills 1 billion people, it will likely kill the entire human population, whereas if it decides not to kill the entire human population, it probably won't hurt anyone). See subsection 4.3.1 for more.

41 This number was selected for illustrative purposes only.

42 For a detailed discussion of this claim, see Pinker, Steven. 2011. *The Better Angels of Our Nature: Why Violence Has Declined*. New York, NY:

Penguin Books. For an account of the randomness of asteroid impacts, see Meier, Matthias, and Sanna Holm-Alwmark. 2017. A Tale of Clusters: No Resolvable Periodicity in the Terrestrial Impact Cratering Record. *Monthly Notices of the Royal Astronomical Society*. URL: https://arxiv.org/pdf/1701.08953.pdf.

43 See Feller, William. 1957. *An Introduction to Probability Theory and Its Applications.* New York, NY: John Wiley & Sons, Inc. Note that this is connected to the "clustering illusion," as well as the "gambler's fallacy" (section 1.6).

44 See Chivian, Eric, and Aaron Bernstein. 2008. *Sustaining Life: How Human Health Depends on Biodiversity.* Oxford: Oxford University Press.

45 Sagan, Carl. 2007. *The Varieties of Scientific Experience: A Personal View of the Search for God.* Edited by Ann Druyan. New York, NY: Penguin Books.

46 It is interesting to note the etymology of the word "organism," which derives from the Greek *organon*, meaning "implement, tool for making or doing." Thus, the sentence "organisms are artifacts" is, from one perspective, virtually tautological. See Online Etymology Dictionary. Accessed March 2, 2017. URL: http://www.etymonline.com/index.php?allowed_in_frame=0&search=organ.

47 Bostrom, Nick. 2012. Existential Risk Prevention as Global Priority. *Global Policy.* 4(1): 15–31.

48 Ibid.

49 See, e.g., Schneider, Susan. 2016. It May Not Feel Like Anything to Be an Alien. *Nautilus.* URL: http://cosmos.nautil.us/feature/72/it-may-not-feel-like-anything-to-be-an-alien.

50 Bostrom, Nick. 2012. Existential Risk Prevention as Global Priority. *Global Policy.* 4(1): 15–31.

51 These categories are elaborations of a four-part scheme originally presented in Bostrom's 2002 article on existential risks. In this scheme, the categories are called "bangs," "crunches," "shrieks," and "whimpers" (in the same order as above). See Bostrom, Nick. 2002. Existential Risks: Analyzing Human Extinction Scenarios and Related Hazards. *Journal of Evolution and Technology.* 9(1).

52 See Ord, Toby. 2015. Will We Cause Our Own Extinction? Natural versus Anthropogenic Existential Risks. CRASSH Cambridge talk. URL: https://www.youtube.com/watch?v=uU0Z4psY32s.

53 As discussed in later sections, agent-tool disasters could be brought

about either by accident or on purpose; that is, through *error or terror*.

54 See Morris, Eric. 2007. From Horse Power to Horsepower. *Access*. URL: http://www.uctc.net/access/30/Access%2030%20-%2002%20-%20 Horse%20Power.pdf.

55 For a list of papers relevant to the simulation argument, go to http://www.simulation-argument.com/.

56 See Bostrom, Nick. 2014. *Superintelligence: Paths, Dangers, Strategies*. Oxford: Oxford University Press.

57 Parfit, Derek. 1984. *Reasons and Persons*. Oxford: Oxford University Press.

58 For more on negative entropy, see the classic Schrödinger, Erwin. 1944/1992. *What Is Life?* Cambridge: Cambridge University Press.

59 For more on the potential good that future technologies could introduce, see Diamandis, Peter. 2014. *Abundance: The Future Is Better Than You Think*. New York, NY: Free Press; and Kurzweil, Ray. 2005. *The Singularity Is Near: When Humans Transcend Biology*. New York, NY: Viking.

60 See Number of Future People. EA Concepts. Accessed on January 14, 2017. URL: https://concepts.effectivealtruism.org/concepts/number-of-future-people/.

61 As for digital immortality through uploading, we should note a potentially crucial difference between (a) *mind uploading*, and (b) *self uploading*. My own view is that, in most cases, uploading technology could enable one to create a "mind clone" by transferring sufficiently precise details about the microstructure of one's brain to computer hardware, the result being a consciousness with an identical past, but different future, than you. This would not in any way transfer the *person*, though, which would be necessary for one to achieve *immortality*. For an insightful discussion of this philosophically convoluted issue, see Schneider, Susan, and Joe Corabi. 2014. The Metaphysics of Uploading. *Journal of Consciousness Studies*. URL: http://schneiderwebsite.com/Susan_Schneiders_Website/Research_files/The%20Metaphysics%20of%20Uploading%20_1.pdf; as well as Chalmers, David. 2010. The Singularity: A Philosophical Analysis. *Journal of Consciousness Studies*. 17: 7–65; and Schneider, Susan. 2008. Future Minds: Transhumanism, Cognitive Enhancement and the Nature of Persons. URL: http://schneiderwebsite.com/uploads/8/3/7/5/83756330/futureminds_transhumanism.pdf.

62 Parfit, Derek. 2011. *On What Matters*. Vol. 2. Oxford: Oxford University Press. Italics added.

63 See also Cotton-Barratt, Owen, and Toby Ord. 2015. Existential Risk and Existential Hope: Definitions. Future of Humanity Institute Technical Report. URL: http://www.fhi.ox.ac.uk/Existential-risk-and-existential-hope.pdf; Beckstead, Nick. 2013. A Proposed Adjustment to the Astronomical Waste Argument. LessWrong. URL: http://lesswrong.com/lw/hjb/a_proposed_adjustment_to_the_astronomical_waste/; and Bostrom, Nick. 2003. Astronomical Waste: The Opportunity Cost of Delayed Technological Development. *Utilitas*. 15(3): 308–314.

64 Bostrom, Nick. 2013. Existential Risk Prevention as Global Priority. *Global Policy*. 4(1): 15–31. Elsewhere, Bostrom writes that the utilitarian imperative to "maximize expected aggregate utility" can be reduced to "minimize existential risk," from which it follows that "the chief goal for utilitarians should be to reduce existential risk." See Bostrom, Nick. 2003. Astronomical Waste: The Opportunity Cost of Delayed Technological Development. *Utilitas*. 15(3): 308–314.

65 See Brand, Stewart. About Long Now. Accessed on March 22, 2017. URL: http://longnow.org/about/.

66 Seth Baum offers a compelling discussion of this issue in a 2014 article about the Toba catastrophe (see section 2.2). To quote him at length: "The Toba eruption may have been the greatest catastrophe in human history. It occurred toward the beginning of the most recent ice age. Ash from the eruption blocked incoming sunlight, causing temperatures to plummet. Living conditions likely became extremely harsh. Some scientists have proposed that the eruption caused a sharp decline in the human population, perhaps to as few as 4,000 people. The human species may have teetered on the edge of extinction. Think about all that's happened in the last 75,000 years. Thousands of generations, billions and billions of people, have lived since then. The agricultural revolution began only about 10,000 years ago, the industrial revolution 250. People today have lives that would be completely inconceivable to the people alive 75,000 years ago. Yet, those people are our ancestors. Without them, none of this would have ever happened. Without them, we would not be here today. We owe our existence to our distant ancestors surviving the Toba eruption." Baum, Seth. 2014. The Lesson of Lake Toba. *Bulletin of the Atomic Scientists*. URL: http://thebulletin.org/lesson-lake-toba7741. (A paragraph break was deleted.)

67 Effective Altruism Foundation. The Importance of the Far Future. Accessed on May 11, 2017. URL: https://ea-foundation.org/blog/the-importance-of-the-far-future/.

68 Matheny, Jason. 2007. Reducing the Risk of Human Extinction. *Risk Analysis*. 27(5): 1335–1344.

69 Ibid. Note that some forms of time discounting are perfectly reasonable. For example, it makes sense to discount money, since money acquired today could be invested to yield even more money in the future. When it comes to discounting *future human lives*, though, serious ethical problems sprout up.

70 See Cowen, Tyler, and Derek Parfit. 1992. Against the Social Discount Rate. In James Fishkin and Peter Laslett (editors), *Justice between Age Groups and Generations*. New Haven, CT: Yale University Press, 144–61.

71 Consider the following blog post from the organization 80,000 Hours, which helpfully elucidates this position: "Which would you choose from these two options? 1. Prevent one person from suffering next year. 2. Prevent 100 people from suffering 100 years from now (the same amount). Most people choose the second option. It's a crude example, but it suggests they value future generations. If people didn't want to leave a legacy to future generations, it's hard to understand why we invest so much in science, create art, and preserve the wilderness. We'd certainly choose the second option. And if you value future generations, then there are powerful arguments that you should focus on helping them." (Some paragraph breaks were deleted.) See 80,000 Hours. 2017. No-One Knew the World's Worst Problem So We Spent 8 Years Trying to Find It. URL: https://80000hours.org/career-guide/world-problems/. Similarly, the Effective Altruism Foundation convincingly argues for the value of future humans as follows: "Future individuals differ from present individuals only in one property—the time they live in. But why should we consider this property to be ethically relevant? From an impartial perspective, the fact that we live in a certain time does not grant this time any special ethical importance. For example, equally intense suffering doesn't feel better or worse depending on the time it is experienced in. If we ground our activism in concern for the wellbeing of others, whether these 'others' live in the distant future, or whether they are suffering presently, should not make a relevant difference. Sentient beings in the future deserve equal moral consideration—disadvantaging them is an unjustified form of discrimination." Effective Altruism Foundation. The Importance of the Far Future. Accessed on May 11, 2017. URL: https://ea-foundation.org/blog/the-importance-of-the-far-future/.

72 See Rawls, John. 1971. *A Theory of Justice*. Cambridge, MA: Belknap Press.

73 All italicized sentences of (1) through (7), as well as the material contained in quotes, are from Bell, Wendell. 1993. Why Should We Care about Future Generations? In Sakae Shimizu et al. *Why Future Generations Now*. Kyoto: Institute for the Integrated Study of Future Generations. For more discussion, see Tonn, Bruce. 2009. Obligations to Future Generations and Acceptable Risks of Human Extinction. *Futures*. 41: 427–435.

74 As Martin Rees puts it, "We cannot confidently guess lifestyles, attitudes, social structures or population sizes a century hence. Indeed, it is not even clear how much longer our descendants would remain distinctively 'human.' Darwin himself noted that 'not one living species will transmit its unaltered likeness to a distant futurity.' Our own species will surely change and diversify faster than any predecessor—via human-induced modifications (whether intelligently controlled or unintended), not by natural selection alone. The post-human era may be only centuries away. And what about Artificial Intelligence? Superintelligent machine could be the last invention that humans need ever make. We should keep our minds open, or at least ajar, to concepts that seem on the fringe of science fiction." See Rees, Martin. 2008. Foreword. In Nick Bostrom and Milan Ćirković (editors), *Global Catastrophic Risks*. Oxford: Oxford University Press.

75 Although there is a small chance that this is not true. For example, astronomers discovered millisecond-long intergalactic pulses of radiation called "fast radio bursts" (FRBs) in 2007. To date, only 17 FRBs have been recorded. These are *extremely odd* signals from the heavens that have no satisfying natural explanation, although possible culprits include "supermassive neutron stars, gamma-ray bursts, and stellar flares." Consequently, two scientists at Harvard recently suggested that FRBs could be the result of extraterrestrials. In particular, an alien civilization could employ FRBs to accelerate spacecraft on interstellar journeys. As they put it, "We envision a beamer that emits the radio waves as a method of launching a light sail. In the same way that a sailboat is pushed by wind, a lightsail is pushed by light and can reach up to the speed of light." The radiation reaching Earth, then, is just the accidental byproduct, or "leakage," from this advanced propulsion technology. Dvorsky, George. 2017. Wild New Theory Suggests Radio Bursts Beyond Our Galaxy Are Powering Alien Starships. *Gizmodo*. URL: http://gizmodo.com/wild-new-theory-suggests-radio-bursts-beyond-our-galaxy-1793130515. See also Lingam, Manasvi, and Abraham Loeb. 2017. Fast Radio Bursts from Extragalactic Light Sails. *The Astrophysical Journal Letters*. URL: https://arxiv.org/pdf/1701.01109.pdf. A few lines

from the relevant paragraph in the body text are excerpted from Torres, Phil. 2016. *The End: What Science and Religion Tell Us about the Apocalypse*. Charlottesville, VA: Pitchstone Publishing.

76 Much of this is borrowed *ad verbum* from SETI. The Drake Equation. Accessed on December 11, 2016. URL: http://www.seti.org/drakeequation.

77 See Brin, David. 1982. The "Great Silence": the Controversy concerning Extraterrestrial Intelligent Life. *Quarterly Journal of the Royal Astronomical Society*. 24: 283–309.

78 This is quoted more or less directly from Hanson, Robin. 1998. The Great Filter—Are We Almost Past It? URL: http://mason.gmu.edu/~rhanson/greatfilter.html. Another concept worth mentioning in the context of posthumanity, the attainment of technological maturity, and the sort of civilization required for a Hansonian "colonization explosion" is the Kardashev scale. This was devised by the Soviet astronomer Nikolai Kardashev in 1964, and it identifies three types of future civilizations based on their capacity to harness energy for the purposes of communication—although today the scale "has been expanded and re-interpreted . . . to include more than just the capacity for communications technology," i.e., scientists "now use [it] to simply describe the amount of energy available to an ETI [extraterrestrial intelligence] for any kind of purpose." In brief, a Type I civilization can use virtually "*all* the power available to it on its home planet" by capturing *in toto* the energy that it receives from its star. Even more, the physicist Michio Kaku states that it can "control earthquakes, the weather—and even volcanoes." (See endnote 10 in chapter 6 for more about this issue from Kaku.) A Type II civilization is one that could "capture the entire energy output of its parent star" through, for example, a Dyson sphere, or a colossal megastructure that encloses a star to harvest nearly all its energy output. Finally, a Type III civilization is one that has colonized its galaxy such that "every scrap of matter—and all its billions of stars—[is] exploited for energy." It would require "anywhere from 100,000 to a million years to transition . . . from a Type II to a Type III" civilization, and "from the perspective of an outside observer, a galaxy occupied by a [Type III civilization] would appear completely invisible, save for the heat leakage which would register in the far infrared." See Dvorsky, George. 2013. How to Measure the Power of Alien Civilizations Using the Kardashev Scale. *iO9*. URL: http://io9.gizmodo.com/5986723/using-the-kardashev-scale-to-measure-the-power-of-extraterrestrial-civilizations. See also Kaku, Michio. 2011. Will Mankind Destroy Itself? Big Think. URL: https://www.youtube.com/watch?v=7NPC47qMJVg.

79 Wilson, E.O. 2006. *Nature Revealed: Selected Writings, 1949–2006*. Baltimore, MD: Johns Hopkins University Press. Similarly, the biologist Ernst Mayr argues in a debate with Carl Sagan that intelligence might be a "lethal mutation," to quote Noam Chomsky's paraphrase. See Mayr, Ernst. 1995. Can SETI Succeed? Not Likely. Planetary Society's Bioastronomy News. 7(3). URL: http://daisy.astro.umass.edu/~mhanner/Lecture_Notes/Sagan-Mayr.pdf; and Chomsky, Noam. 2010. Human intelligence and the environment. Accessed on February 27, 2017. URL: https://chomsky.info/20100930/.

80 Indeed, Hanson's nine-part distinction could be too simple: there might exist many more evolutionary transitions at which a Great Filter could reside. Alternatively, A.H. Knoll and Richard Bambach argue that the history of life on Earth consists (so far) of six "megatrajectories," where "each megatrajectory adds new and qualitatively distinct dimensions to the way life utilizes ecospace." These megatrajectories are: (1) the origin of life to the Last Common Ancestor, (2) prokaryote diversification, (3) unicellular eukaryote diversification, (4) multicellular organisms, (5) land organisms, and (6) appearance of intelligence and technology. Knoll, A.H., and Richard Bambach. 2000. Directionality in the History of Life: Diffusion from the Left Wall or Repeated Scaling of the Right." *Paleobiology*. 26(4): 1–14. The scholars Milan Ćirković and Robert Bradbury further explore the possibility of "post-biological evolution," i.e., evolution after humanity attains a posthuman state, perhaps involving advanced artificial intelligences, as constituting a seventh megatrajectory. See Ćirković, Milan, and Robert Bradbury. 2006. Galactic Gradients, Postbiological Evolution and the Apparent Failure of SETI. *New Astronomy*. 11(8): 628–639.

81 Hanson, Robin. 1998. The Great Filter—Are We Almost Past It? URL: http://mason.gmu.edu/~rhanson/greatfilter.html. See also Hanson, Robin. The Great Filter. TEDxLimassol. URL: https://www.youtube.com/watch?v=aspMV6ERqpo.

82 The same goes for generating life from non-life in the laboratory.

83 Bostrom, Nick. 2008. Where Are They? Why I Hope the Search for Extraterrestrial Life Finds Nothing. *MIT Technology Review*. May/June: 72–77.

84 Ibid.

85 The Hubble volume is, to quote Bostrom, "the part of the expanding universe that is theoretically accessible from where we are now." From Bostrom, Nick. 2014. *Superintelligence: Paths, Dangers, Strategies*. Oxford: Oxford University Press.

86 Tonn, Bruce, and Dorian Stiefel. 2012. The Race for Evolutionary Success. *Sustainability*. 4: 1787–1805.

87 See Tversky, Amos, and Daniel Kahneman. 1983. Extensional versus Intuitive Reasoning: The conjunction fallacy in probability judgment. *Psychological Review*. 90(4): 293–315.

88 Even if one assigns a minuscule 0.001 probability to Linda being a bank teller and a whopping 0.999 probability of her being a feminist, the probability of these *together* is still less than the probability of her being a bank teller *alone*. To calculate this, we use $P(A)$ x $P(B)$ = $P(A$ and $B)$, assuming independence. Plugging in the relevant numbers, we get a *joint probability* of 0.000999, which is less than 0.001.

89 As I have written elsewhere, this is why theism by itself is necessarily more probable than any religious system built up around it, many of which are extremely elaborate, thus making them far less probable than theism, which some scholars argue is already quite improbable.

90 For more, see David, Marian. 2016. The Correspondence Theory of Truth. *The Stanford Encyclopedia of Philosophy*. URL: https://plato.stanford.edu/archives/fall2016/entries/truth-correspondence/.

91 Or, put differently, the more evidence will be required to justify the theory. See Oddie, Graham. 2016. Truthlikeness. *The Stanford Encyclopedia of Philosophy*. URL: https://plato.stanford.edu/archives/win2016/entries/truthlikeness/.

92 The probability of A *or* B occurring is the probability of each added together minus the probability of both occurring together, or $P(A)$ + $P(B)$ - $P(A$ and $B)$ = $P(A$ or $B)$. Thus, using the probabilities of endnote 95, we get 1 (from 0.999 + 0.001) - 0.000999 (from 0.999 x 0.001), which equals 0.999001. And, of course, 0.999001 is greater than 0.999, or the probability that Linda is active in the feminist movement. It follows that (b) is more probable than (a).

93 Ćirković, Milan. 2008. Observation Selection Effects and Global Catastrophic Risks. In Nick Bostrom and Milan Ćirković (editors), *Global Catastrophic Risks*. Oxford: Oxford University Press. See also Ćirković, Milan, Anders Sandberg, and Nick Bostrom. 2010. Anthropic Shadow: Observation Selection Effects and Human Extinction Risks. *Risk Analysis*. 30(10): 1495–1506; and Bostrom, Nick. 2005. Self-Location and Observation Selection Theory: An Advanced Introduction. URL: http://www.anthropic-principle.com/preprints/self-location.html.

94 Gray, Peter. 2011. *Psychology*. New York, NY: Worth Publishers.

95 Wilke, Andreas, and Rui Mata. 2012. Cognitive Bias. In V.S. Ramachan-

dran (editor), *The Encyclopedia of Human Behavior*. Amsterdam: Academic Press.

96 Bostrom, Nick. 2002. Existential Risks: Analyzing Human Extinction Scenarios and Related Hazards. *Journal of Evolution and Technology*. 9(1).

97 Yudkowsky, Eliezer. 2008. Cognitive Biases Potentially Affecting Judgement of Global Risks. In Nick Bostrom and Milan Ćirković (editors), *Global Catastrophic Risks*. Oxford: Oxford University Press.

98 Begley, Sharon. 2009. Why We Believe Lies, Even When We Learn the Truth. *Newsweek*. URL: http://www.newsweek.com/why-we-believe-lies-even-when-we-learn-truth-78775.

99 Quoted from Scope Neglect. Wikipedia. Accessed on January 4, 2017. URL: https://en.wikipedia.org/wiki/Scope_neglect.

100 Scicurious. 2013. The Superiority Illusion: Where Everyone Is Above Average. *Scientific American*. URL: https://blogs.scientificamerican.com/scicurious-brain/the-superiority-illusion-where-everyone-is-above-average/.

101 Sharot, Tali. 2012. *The Optimism Bias: A Tour of the Irrationally Positive Brain*. New York, NY: Vintage Books.

102 Anchoring. Wikipedia. Accessed on March 27, 2017. URL: https://en.wikipedia.org/wiki/Anchoring.

103 Bar-Hillel, Maya. 1980. The Base-Rate Fallacy in Probability Judgments. *Acta Psychologica*. 44(3): 211–233. Italics added.

104 Wilke, Andreas, and Rui Mata. 2012. Cognitive Bias. In V.S. Ramachandran (editor), *The Encyclopedia of Human Behavior*. Amsterdam: Academic Press. A single comma was added.

105 See Overconfidence Effect. Psychology Concepts. Accessed on February 2, 2017. URL: http://www.psychologyconcepts.com/overconfidence-effect/.

106 For more, see Yudkowsky, Eliezer. 2008. Cognitive Biases Potentially Affecting Judgement of Global Risks. In Nick Bostrom and Milan Ćirković (editors), *Global Catastrophic Risks*. Oxford: Oxford University Press.

107 Parts of this section are excerpted from Torres, Phil. 2016. The Clash of Eschatologies: The Role of End-Times Thinking in World History. *Skeptic*. URL: https://goo.gl/3HhvrB.

108 Note that the word "rapture" does not appear anywhere in the Bible.

109 Dispensationalists like Tim LaHaye (coauthor of the *Left Behind* series)

and John Hagee (who heads the powerful Zionist organization called Christians United for Israel) disagree about which of these institutions the Antichrist will take over.

110 Although many Christians today believe that Revelation is about future rather than past events—a theological position called "futurism"—this book, written by John of Patmos, was almost certainly about the particular historical period in which he lived. For example, consider the "Number of the Beast," which is traditionally understood to be 666, although an ancient fragment of Revelation called Papyrus 115 identifies it as 616. As Frances Flannery observes, "The infamous number '666' . . . indicates that John had the [Roman] Emperors Nero or Domitian in mind when he wrote about the first beast. In the Jewish tradition of *gematria*, numbers were also codes for words, since Hebrew letters possess numerical equivalents. The author 'John,' most likely a Palestinian Jewish Christian, draws on this Jewish tradition when he writes that 'the number of the name' (Rev. 13:17) is a code for a name. The obvious candidate for the name is Nero Caesar, which yields '666' when translated from Greek into Hebrew. In fact, some early manuscripts of the New Testament list '616' as the number of the beast because the Latin form of 'Nero Caesar' transliterated into Hebrew is spelled *nrw qsr*, which adds up to 616." Flannery, Frances. 2016. *Understanding Apocalyptic Terrorism: Countering the Radical Mindset.* New York, NY: Routledge.

111 Note that some dispensationalists distinguish between the Gog and Magog mentioned in Ezekiel and the one being mentioned in Revelation. These are two distinct entities. For clarification, see Got Questions? What are Gog and Magog? Accessed on April 11, 2017. URL: https://www.gotquestions.org/Gog-Magog.html.

112 Despite the fact that Constantinople was already conquered by the Ottoman Turks in 1453!

113 For a helpful overview of one version of Islamic eschatology, see Qadhi, Yasir. 2011. The Mahdi between Fact and Fiction. YouTube. URL: https://www.youtube.com/watch?v=1oCf7ae__kk.

114 Pew Research Center. 2010. Jesus Christ's Return to Earth. URL: http://www.pewresearch.org/fact-tank/2010/07/14/jesus-christs-return-to-earth/. See also See Pew Research Center. 2013. U.S. Christians' Views on the Return of Christ. URL: http://www.pewforum.org/2013/03/26/us-christians-views-on-the-return-of-christ/.

115 Pew Research Center. 2012. The World's Muslims: Unity and Diversity; Chapter 3: Articles of Faith. URL: http://www.pewforum.org/2012/08/09/the-worlds-muslims-unity-and-diversity-3-articles-of-faith/.

116 See McKnight, Scot. 2005. *Jesus and His Death: Historiography, the Historical Jesus, and Atonement Theory*. Waco, TX: Baylor University Press; and Schweitzer, Albert. 1911. *The Quest of the Historical Jesus*. London: Adam and Charles Black.

117 Fromherz, Allen. 2009. Final, Judgment. In John Esposito (editor), *The Oxford Encyclopedia of the Islamic World*. John Esposito. Oxford: Oxford University Press.

118 See Torres, Phil. 2016. The Clash of Eschatologies: The Role of End-Times Thinking in World History. *Skeptic*. URL: https://goo.gl/3HhvrB; and Huntington, Samuel P. 1996. *The Clash of Civilizations and the Remaking of World Order*. New York, NY: Touchstone.

119 For more, see Landes, Richard. 2011. *Heaven on Earth: The Varieties of the Millennial Experience*. Oxford: Oxford University Press; Flannery, Frances. 2016. *Understanding Apocalyptic Terrorism: Countering the Radical Mindset*. New York, NY: Routledge; and Torres, Phil. 2016. *The End: What Science and Religion Tell Us about the Apocalypse*. Charlottesville, VA: Pitchstone Publishing.

120 See, for example, Boyer, Paul. 2003. When U.S. Foreign Policy Meets Biblical Prophecy. *AlterNet*. URL: http://www.alternet.org/story/15221/when_u.s._foreign_policy_meets_biblical_prophecy; Campolo, Tony. 2005. The Ideological Roots of Christian Zionism. *Tikkun*. 20:1; and Haija, Rammy. 2006. The Armageddon Lobby: Dispensationalist Christian Zionism and the Shaping of US Policy Towards Israel-Palestine. *Holy Land Studies*. 5(1): 75–95.

121 McCants, Will. 2015. *The ISIS Apocalypse: The History, Strategy, and Doomsday Vision of the Islamic State*. New York, NY: St. Martin's Press.

122 See Chirot, Daniel, and Clark McCauley. 2006. *Why Not Kill Them All? The Logic and Prevention of Mass Political Murder*. Princeton, NJ: Princeton University Press.

123 Ibid.

124 Flannery, Frances. 2016. *Understanding Apocalyptic Terrorism: Countering the Radical Mindset*. New York, NY: Routledge.

125 As the sociologist James Hughes puts a similar point, "The millennial impulse is ancient and universal in human culture and is found in many contemporary, purportedly secular and scientific, expectations about the future." Yet he adds that "millennialist responses are inevitable in the consideration of potential catastrophic risks and are not altogether unwelcome. Secular techno-millennials and techno-apocalyptics can play critical roles in pushing reluctant institutions towards positive social

change or to enact prophylactic policies just as religious millennialists have in the past. But the power of millennialism comes with large risks and potential cognitive errors which require vigilant self-interrogation to avoid." Hughes, James. 2008. Millennial Tendencies in Responses to Apocalyptic Threats. In Nick Bostrom and Milan Ćirković (editors), *Global Catastrophic Risks*. Oxford: Oxford University Press.

126 That is, not anecdotal or subjective evidence, but evidence that can be *checked and double checked* by different people in different places at different times. Science only cares about checkable evidence.

127 This claim is closely tied to the confirmation bias discussed in the previous section.

128 It is helpful to recall here that beliefs must always be the *destinations* of a journey guided by evidence and never the *points of departure*.

129 As I explore in a previous book, bracketing the Gettier problem, knowledge consists of beliefs that are both true and justified. This is the Platonic "justified true beliefs" analysis. In contrast, faith—at least in the *epistemic* sense, as there are other senses of the term, such as the *fiducial* sense—refers to beliefs that are always unjustified and that could be either true or false. Thus, since faith-based beliefs lack justification, one should always avoid them. Once again, this gestures at why religious vaticinations ought to be ignored and shunned while scientific doomsayers should be taken seriously. For a more detailed discussion of these issues, see appendix 4 of Torres, Phil. *The End: What Science and Religion Tell Us about the Apocalypse*. Charlottesville, VA: Pitchstone Publishing.

Chapter 2

1 See Tyson, Neil deGrasse. 2013. The Universe Is Trying to Kill You. Big Think. URL: https://www.youtube.com/watch?v=Fw62e4SDHHo.

2 Parts of this section have been excerpted from Torres, Phil. 2016. *The End: What Science and Religion Tell Us about the Apocalypse*. Charlottesville, VA: Pitchstone Publishing. See also Evans, Robert. 2002. Blast from the Past. *Smithsonian*. URL: http://www.smithsonianmag.com/history/blast-from-the-past-65102374/?no-, and Raffles, Sophia. 1830. *Memoir of the Life and Public Services of Sir Thomas Stamford Raffles, F.R.S. &c., Particularly in the Government of Java 1811–1816, and of Bencoolen and Its Dependencies 1817–1824: With Details of the Commerce and Resources of the Eastern Archipelago, and Selections from his Correspondence*. London: John Murray (publisher).

3   See Pirages, Dennis. Nature, Disease, and Globalization: An Evolution-
    ary Perspective. In George Modelski, Tessaleno Devezas, and William
    Thompson (editors), *Globalization as Evolutionary Process: Modeling
    Global Change*. New York, NY: Routledge; Evans, Robert. 2002. Blast
    from the Past. *Smithsonian*. URL: http://www.smithsonianmag.com/
    history/blast-from-the-past-65102374/?no-; and Wood, D'Arcy Gillen.
    2014. The Volcano That Changed the Course of History. *Slate*. URL:
    http://www.slate.com/articles/health_and_science/science/2014/04/
    tambora_eruption_caused_the_year_without_a_summer_cholera_
    opium_famine_and.html.

4   Evans, Robert. 2002. Blast from the Past. *Smithsonian*. URL: http://www.
    smithsonianmag.com/history/blast-from-the-past-65102374/?no-.

5   Consider, e.g., a line from *Frankenstein* in which Victor Frankenstein
    describes a nighttime storm: "Vivid flashes of lightning dazzled my eyes,
    illuminating the lake, making it appear like a vast sheet of fire; then for
    an instant everything seemed of a pitchy darkness, until the eye recov-
    ered itself from the preceding flash." Compare this to an excerpt from
    a letter that Shelley wrote to her sister: "One night we enjoyed a finer
    storm than I had ever before beheld. The lake was lit up—the pines on
    Jura made visible and all the scene illuminated for an instant, when a
    pitchy blackness succeeded, and the thunder came in frightful bursts
    over our heads amid the darkness." See Greenfieldboyce, Nell. 2007.
    Did Climate Inspire the Birth of a Monster? *NPR*. URL: http://www.npr.
    org/2007/08/13/12688403/did-climate-inspire-the-birth-of-a-monster;
    and Smothers, Richard. 1984. The Great Tambora Eruption in 1815 and
    Its Aftermath. *Science*. 224(4654).

6   Rampino, Michael. 2008. Super-volcanism and Other Geophysical Pro-
    cesses of Catastrophic Import. In Nick Bostrom and Milan Ćirković
    (editors), *Global Catastrophic Risks*. Oxford: Oxford University Press.

7   See, e.g., Mason, Ben, David Pyle, and Clive Oppenheimer. 2004. The
    Size and Frequency of the Largest Explosive Eruptions on Earth. *Bul-
    letin of Volcanology*. 66(8): 735–748.

8   Rampino, Michael. 2008. Super-volcanism and Other Geophysical Pro-
    cesses of Catastrophic Import. In Nick Bostrom and Milan Ćirković
    (editors), *Global Catastrophic Risks*. Oxford: Oxford University Press.

9   Britt, Robery Roy. 2005. Super Volcano Will Challenge Civilization,
    Geologists Warn. *Live Science*. URL: http://www.livescience.com/200-
    super-volcano-challenge-civilization-geologists-warn.html.

10 Rampino, Michael. 2008. Super-volcanism and other geophysical pro-

cesses of catastrophic import. In Nick Bostrom and Milan Ćirković (editors), *Global Catastrophic Risks*. Oxford: Oxford University Press.

11   Geological Society. Super-eruptions: global effects and future threats. URL: https://www.geolsoc.org.uk/~/media/shared/documents/education%20and%20careers/Super_eruptions.pdf?la.

12   Rampino, Michael. 2008. Super-volcanism and Other Geophysical Processes of Catastrophic Import. In Nick Bostrom and Milan Ćirković (editors), *Global Catastrophic Risks*. Oxford: Oxford University Press. Some citations originally included in this quote have been removed.

13   Ibid.

14   See Yellowstone National Park. Supervolcano under Yellowstone Is Twice as Large as Previously Thought. Accessed on December 13, 2016. URL: http://www.yellowstonepark.com/super-volcano-twice-large/.

15   Online Report of a Geological Society Working Group. 2005. Executive Summary. URL: https://www.google.com/url?sa=t&rct=j&q=&esrc=s&source=web&cd=2&ved=0ahUKEwikouSdlPLQAhXD5CYKHQkfBRMQFgghMAE&url=http%3A%2F%2Fwww.bioelectricalwellness.com%2Fdocuments%2FVolcanos%2520-%2520Super-Eruptions.doc&usg=AFQjCNEh0yigyVXLjtnkUXpLdfIW62WzcQ&sig2=izqhfFqIQZx5AWCO8tnF6w.

16   Baum, Seth. 2015. The Far Future Argument for Confronting Catastrophic Threats to Humanity: Practical Significance and Alternatives. *Futures*. 72: 86–96.

17   Ibid.

18   See Kilbourne, Edwin. 2008. Plagues and Pandemics: Past, Present, and Future. In Nick Bostrom and Milan Ćirković (editors), *Global Catastrophic Risks*. Oxford: Oxford University Press.

19   For details on this topic, see Cohen, Mark. 1989. *Health and the Rise of Civilization*. New Haven, CT: Yale University Press.

20   Kilbourne, Edwin. 2008. Plagues and Pandemics: Past, Present, and Future. In Nick Bostrom and Milan Ćirković (editors), *Global Catastrophic Risks*. Oxford: Oxford University Press.

21   Ibid.

22   See World Health Organization. 2015. Fact Sheet: World Malaria Report 2015. URL: http://www.who.int/malaria/media/world-malaria-report-2015/en/.

23   See Global Challenges Foundation. 2015. 12 Risks that Threaten Human

Civilisation. URL: https://api.globalchallenges.org/static/wp-content/uploads/12-Risks-with-infinite-impact.pdf. Italics added. Note also that a pandemic is simply an epidemic that spreads around the world, with "pan" meaning as "all."

24 See Senthilingam, Meera. 2017. Seven Reasons We're at More Risk than Ever of a Global Pandemic. CNN. URL: http://www.cnn.com/2017/04/03/health/pandemic-risk-virus-bacteria/index.html.

25 Ibid.

26 This is an issue that overlaps with the topic of "agential error," discussed in subsection 4.3.2.

27 See Kilbourne, Edwin. 2008. Plagues and Pandemics: Past, Present, and Future. In Nick Bostrom and Milan Ćirković (editors), *Global Catastrophic Risks*. Oxford: Oxford University Press.

28 See James, John. 2013. A New, Evidence-Based Estimate of Patient Harms Associated with Hospital Care. *Journal of Patient Safety*. 9(3): 122–128.

29 See Siegel, Rebecca, Kimberly Miller, and Ahmedin Jemal. 2016. Cancer Statistics, 2016. *CA: A Cancer Journal for Clinicians*. 66(1): 7–30.

30 See Miller, Kelli. 2015. Superbugs: What They Are and How You Get Them. WebMD. URL: http://www.webmd.com/a-to-z-guides/news/20150417/superbugs-what-they-are#1.

31 Quoted in ibid.

32 Quoted in Fox, Maggie. 2016. Drug-Resistant Superbugs Are a "Fundamental Threat," WHO Says. NBC News. URL: http://www.nbcnews.com/health/health-news/who-labels-drug-resistant-superbugs-fundamental-threat-humans-n651981. Also note that the risks associated with superbugs could be classified as an unintended consequence (see chapter 3), since they are the inadvertent outcome of large-scale purposive human action. Indeed, one of the major causes of superbugs is the use of antibiotics in food-producing livestock. This "has contributed to the rise of antibiotic resistant bacteria in animals, which are then transferred to humans when we eat foods from these animals like meat or milk." Another factor is the overuse of antibiotics among humans; an estimated one-third of antibiotic prescriptions are unnecessary, according to the CDC. But the central driver of resistance to antibiotics is the Darwinian mechanism of natural selection, whereby genes less suited to their selective environment are probabilistically removed from the population. In other words, bacteria are evolving to resist the drugs that we put into their milieus, and this naturalistic factor nudges superbugs

into the category of natural risks. It is nature's tinkering with DNA that enables the formation of treatment-resistant germs. See Kane, Michaela. 2016. "Superbugs" and the Very Real Threat of Untreatable Infections. *Harvard Health Publications*. URL: http://www.health.harvard.edu/blog/superbugs-real-rise-antibiotic-resistant-bacteria-201607069856.

33 See Gupta, Sanjay. 2017. The Big One Is Coming, and It's Going to Be a Flu Pandemic. CNN. URL: http://www.cnn.com/2017/04/07/health/flu-pandemic-sanjay-gupta/index.html.

34 In other words, as I have previously written, "some dinosaurs survived and evolved into birds. So, next time you have a chicken sandwich, you can just as accurately call it a 'dinosaur sandwich.'" See Torres, Phil. 2016. 10 Catastrophes That Could Cause the Extinction of Our Species. *Patheos*. URL: http://www.patheos.com/blogs/friendlyatheist/2016/07/18/10-catastrophes-that-could-cause-the-extinction-of-our-species/.

35 See Endangered Species International. The Five Worst Mass Extinctions. Accessed on December 13, 2016. URL: http://www.endangered-speciesinternational.org/overview.html.

36 See NASA. 2017. Near Earth Object Program. URL: http://neo.jpl.nasa.gov/orbits/.

37 Napier, William. 2008. Hazards from Comets and Asteroids. In Nick Bostrom and Milan Ćirković (editors), *Global Catastrophic Risks*. Oxford: Oxford University Press.

38 Ibid.

39 Ibid. A citation original to this quote has been removed.

40 Ibid.

41 This section is indebted to Dar, Anon. 2008. Influence of Supernovae, Gamma-Ray Bursts, Solar Flares, and Cosmic Rays on the Terrestrial Environment. In Nick Bostrom and Milan Ćirković (editors), *Global Catastrophic Risks*. Oxford: Oxford University Press.

42 See Häggström, Olle. 2016. *Here Be Dragons: Science, Technology and the Future of Humanity*. Oxford: Oxford University Press.

43 Dar, Arnon. 2008. Influence of Supernovae, Gamma-Ray Bursts, Solar Flares, and Cosmic Rays on the Terrestrial Environment. In Nick Bostrom and Milan Ćirković (editors), *Global Catastrophic Risks*. Oxford: Oxford University Press.

44 Ibid.

45 Ibid.

46   It follows directly from this that the big bang literally happened *every-where.*

47   Google definition. Accessed on March 19, 2017. URL: https://www.google.com/search?q=entropy&oq=entropy&aqs=chrome..69i57j0l5.656j0j4&sourceid=chrome&ie=UTF-8#dobs=entropy.

48   See Overbye, Dennis. 2008. Kissing the Earth Goodbye in About 7.59 Billion Years. *New York Times.* URL: http://www.nytimes.com/2008/03/11/science/space/11earth.html.

49   See Adams, Fred. 2008. Long-Term Astrophysical Processes. In Nick Bostrom and Milan Ćirković (editors), *Global Catastrophic Risks.* Oxford: Oxford University Press. Unlike the other risks discussed here, which appear quite unlikely, the entropy death of the universe looks to be quite certain!

50   Ibid. Although this doesn't mean that we would survive it.

51   See Kaku, Michio. 2005. *Parallel Worlds: A Journey through Creation, Higher Dimensions, and the Future of the Cosmos.* New York, NY: Doubleday.

52   See Torres, Phil. 2016. The Clash of Eschatologies: The Role of End-Times Thinking in World History. *Skeptic.* URL: https://goo.gl/b4rBbR.

Chapter 3

1   Merton, Robert. 1936. The Unanticipated Consequences of Purposive Social Action. *American Sociological Review.* 1(6): 894–904.

2   Indeed, insofar as existential risk scholars make predictions about future catastrophes, it is *precisely* to ensure that such catastrophes never occur—a kind of *intended* self-defeating prophecy. See the last paragraph of the postscript.

3   Winner, Langdon. 1977. *Autonomous Technology: Technics-out-of-Control as a Theme in Political Thought.* Cambridge, MA: MIT Press.

4   Jacquet, Jennifer. 2017. The Anthropocene. Edge.org. URL: https://www.edge.org/response-detail/27096.

5   To be clear, "climate change" can refer to any instance of "a change in the statistical distribution of weather patterns when that change lasts for an extended period of time" (to quote Wikipedia), whether caused by humans or natural phenomena. The definition presented here is contextualized to our present anthropocenic moment.

6   While there is considerable uncertainty surrounding the possibility of out-of-control warming, as noted below, scientists generally see this scenario as improbable.

7   See Billings, Lee. 2013. Fact or Fiction? We Can Push the Planet into a Runaway Greenhouse Apocalypse. *Scientific American*. URL: https://www.scientificamerican.com/article/fact-or-fiction-runaway-greenhouse/.

8   See Emerging Technology from the arXiv. 2012. How Likely Is a Runaway Greenhouse Effect on Earth? *The MIT Review*. URL: https://www.technologyreview.com/s/426608/how-likely-is-a-runaway-greenhouse-effect-on-earth/.

9   In game theoretic terms, the tragedy of the commons is a multi-player prisoner's dilemma: it is in each person's self-interest to "defect"—i.e., *not* to cooperate with the whole by, instead, adding another animal to one's herd. Here, defecting *strongly dominates* cooperating, meaning that all rational players will choose the former, even though doing this results in a bad outcome for everyone.

10 Hardin, Garrett. 1968. The Tragedy of the Commons. *Science*. 162: 1243–1248. We should also note here what is called "Hardin's First Law of Human Ecology." This directly addresses the problem of unintended consequences, asserting that "we can never do merely one thing. Any intrusion into nature has numerous effects, many of which are unpredictable."

11 Furthermore, economic competition offers overwhelming near-term incentives for companies and governments alike to continue emitting carbon dioxide; indeed, Lord Nicholas Stern, in the influential "Stern Review," claims that climate change "is the greatest market failure the world has ever seen." Quoted in Rotman, David. 2011. Nicholas Stern. *MIT Technology Review*. URL: https://www.technologyreview.com/s/424380/nicholas-stern/. The activist Naomi Klein also tackles this tangled conundrum in her book, Klein, Naomi. 2014. *This Changes Everything: Capitalism vs. the Climate*. New York, NY: Simon & Schuster.

12 Persson, Ingmar, and Julian Savulescu. 2012. *Unfit for the Future: The Need for Moral Enhancement*. Oxford: Oxford University Press. As we will discuss later on, Persson and Savulescu believe that a combination of cognitive and moral enhancements could lift humanity out of the labyrinth of the commons tragedy, and thus that the use of both should become widespread in contemporary society.

13 See IPCC. 2014. Climate Change 2014, *Synthesis Report*. URL: https://www.ipcc.ch/news_and_events/docs/ar5/ar5_syr_headlines_en.pdf.

14 A similar point is made in Torres, Phil. 2016. We're Speeding toward a Climate Change Catastrophe—and That Makes 2016 the Most Important Election in a Generation. *Salon*. URL: http://www.salon.com/2016/04/10/were_speeding_toward_a_climate_change_catastrophe_and_that_makes_2016_the_most_important_election_in_a_generation/.

15 That is, in the absence of some magical new technology that enables scientists to quickly and effectively restore pre-industrial levels of atmospheric carbon dioxide. This could very well be developed in the future, but we should not count on such a breakthrough. Indeed, doing so would be like smoking a pack of cigarettes every day on the assumption that by the time one gets cancer, a safe and effective cure will have been found. See Clark, Peter et al. 2016. Consequences of Twenty-First-Century Policy for Multi-millennial Climate and Sea-Level Change. *Nature Climate Change*. 6: 360–369.

16 See, in this order: Romps, David et al. 2014. Projected Increase in Lightning Strikes in the United States due to Global Warming. *Science*. 346(6211): 851–854; Sarfaty, Mona. 2016. Making the Connection: Climate Change, Allergies, and Asthma. URL: http://medsocietiesforclimatehealth.org/wp-content/uploads/2016/10/2016-APHA-Allergy-Asthma-MS.pptx; and Mitrovica, Jerry et al. 2015. Reconciling Past Changes in Earth's Rotation with 20th Century Global Sea-Level Rise: Resolving Munk's Enigma. *Science Advances*. 1(11).

17 See Sherwood, Steven, and Matthew Huber. 2010. An Adaptability Limit to Climate Change due to Heat Stress. *Proceedings of the National Academy of Sciences*. 107(21): 9552–9555; and Buzan, Jonathan, Keith Oleson, and Matthew Huber. 2015. Implementation and Comparison of a Suite of Heat Stress Metrics within the Community Land Model Version 4.5. *Geoscientific Model Development*. 8: 151–170.

18 See NOAA. 2015. Global Analysis—Annual 2015. URL: https://www.ncdc.noaa.gov/sotc/global/201513.

19 Thompson, Andrea. 2017. 2016 Was the Hottest Year on Record. *Scientific American*. URL: https://www.scientificamerican.com/article/2016-was-the-hottest-year-on-record/.

20 Quoted in Sinnett, Danielle, Nick Smith, and Sarah Burgess (editors). 2015. *Handbook on Green Infrastructure: Planning, Design and Implementation*. Cheltenham, U.K.: Edward Elgar Publishing Limited.

21   Parts of this discussion are excerpted from Torres, Phil. 2016. Biodiversity Loss: An Existential Risk Comparable to Climate Change. *Bulletin of the Atomic Scientists*. URL: http://thebulletin.org/biodiversity-loss-existential-risk-comparable-climate-change9329; Torres, Phil. 2016. Biodiversity Loss and the Doomsday Clock: An Invisible Disaster Almost No One Is Talking About. *Common Dreams*. URL: http://www.commondreams.org/views/2016/02/10/biodiversity-loss-and-doomsday-clock-invisible-disaster-almost-no-one-talking-about; and several other articles that I have published.

22   See Convention on Biological Diversity. 2010. Global Biodiversity Outlook 3. URL: https://www.cbd.int/doc/publications/gbo/gbo3-final-en.pdf. Incidentally, people often fail to consider how human activity might affect plant life, a phenomenon that scientists call "plant blindness."

23   World Wide Fund for Nature. 2016. Living Planet Report 2016. URL: http://awsassets.panda.org/downloads/lpr_living_planet_report_2016_summary.pdf. A related concept is called "Earth Overshoot Day." This refers to the point each year when the total human consumption of resources thus far exceeds Earth's capacity to regenerate those resources. In 1990, Earth Overshoot Day was December 7; in 2000, it was November 1; in 2010, it was August 21; and by 2016, it was August 8. See Wikipedia. Earth Overshoot Day. Accessed on April 2, 2017. URL: https://en.wikipedia.org/wiki/Earth_Overshoot_Day.

24   World Wide Fund for Nature. 2016. Living Planet Report 2016. URL: http://awsassets.panda.org/downloads/lpr_living_planet_report_2016_summary.pdf.

25   See Vidal, John. 2013. One in Five Reptile Species Faces Extinction—Study. *Guardian*. URL: https://www.theguardian.com/environment/2013/feb/15/reptile-species-face-extinction; and Zimmer, Carl. 2017. Most Primate Species Threatened with Extinction, Scientists Find. *New York Times*. URL: https://www.nytimes.com/2017/01/18/science/almost-two-thirds-of-primate-species-near-extinction-scientists-find.html?_r=0.

26   Along these lines, the Cavendish banana sold in grocery stores today is facing extinction due to a fungal infection. If it were to go extinct, it wouldn't be the first time something like this happened. Until the 1950s, the primary banana sold in Europe and the Americas was the Gros Michel, or "Big Mike," which is said to have had a superior taste to the bananas we now consume. (Its peel was also, apparently, much slipperier, which explains the slip-on-a-banana-peel gag in old slapstick comedies.) But due to an outbreak of Panama disease, caused by the

fungus *Fusarium*, Gros Michel crops were decimated and banana plantations abandoned. The very same thing could happen to the Cavendish banana. See Torres, Phil. 2016. Biodiversity Loss and the Doomsday Clock: An Invisible Disaster Almost No One Is Talking About. *Common Dreams*. URL: http://www.commondreams.org/views/2016/02/10/biodiversity-loss-and-doomsday-clock-invisible-disaster-almost-no-one-talking-about.

27 Renee Cho, Losing Our Coral Reefs. State of the Planet. URL: http://blogs.ei.columbia.edu/2011/06/13/losing-our-coral-reefs/.

28 Connor, Steve. 2015. Ocean Acidification Killed Off More Than 90 Per cent of Marine Life 252 Million Years Ago, Scientists Believe. *Independent*. URL: http://www.independent.co.uk/news/science/ocean-acidification-killed-off-more-than-90-per-cent-of-marine-life-252-million-years-ago-scientists-10165989.html.

29 Hand, Eric. 2015. Acid Oceans Cited in Earth's Worst Die-Off. *Science*. 348(6231): 165–166. Italics added.

30 Ibid.

31 Quoted in ibid. Or, to be more precise, evolution occurs rapidly on geological timescales, while occurring imperceptibly in realtime. Note also that "the amount of methane in the atmosphere has increased from 0.7 parts per million to 1.7 parts per million." This is a very ominous trend, indeed. As for the rate of evolution, it's worth noting that, "the rate of evolution trails the rate of climate change by a factor of 10,000." See Jamail, Dahr. 2013. What These Climate Scientists Said About Earth's Future Will Terrify You. *Mother Jones*. URL: http://www.motherjones.com/environment/2013/12/climate-scientist-environment-apocalypse-human-extinction.

32 Connor, Steve. 2015. Ocean Acidification Killed Off More Than 90 Per cent of Marine Life 252 Million Years Ago, Scientists Believe. *Independent*. URL: http://www.independent.co.uk/news/science/ocean-acidification-killed-off-more-than-90-per-cent-of-marine-life-252-million-years-ago-scientists-10165989.html. Italics added. Furthermore, ocean acidification is becoming so pronounced that the shells of "tiny marine snails that live along North America's western coast" are literally dissolving in the water, resulting in "pitted textures" that give the shells a "cauliflower" or "sandpaper" appearance.See Kintisch, Eli. 2014. Snails Are Dissolving in Pacific Ocean. *Science*. URL: http://www.sciencemag.org/news/2014/05/snails-are-dissolving-pacific-ocean.

33 See Torres, Phil. 2016. Biodiversity Loss and the Doomsday Clock: An

Invisible Disaster Almost No One Is Talking About. *Common Dreams.* URL: http://www.commondreams.org/views/2016/02/10/biodiversity-loss-and-doomsday-clock-invisible-disaster-almost-no-one-talking-about.

34 See Asia-Pacific Correspondent, Daniel Howden, and Kathy Marks. 2008. The World's Rubbish Dump: A Tip That Stretches from Hawaii to Japan. *The Independent.* URL: http://www.independent.co.uk/environment/green-living/the-worlds-rubbish-dump-a-tip-that-stretches-from-hawaii-to-japan-778016.html.

35 A population can, of course, decline without this resulting in extinction. We have so far been considering only the former.

36 Ceballos, Gerardo et al. 2015. Accelerated Modern Human-Induced Species Losses: Entering the Sixth Mass Extinction. *Science Advances.* 1(5).

37 See Jacquet, Jennifer. 2016. Human Error: Survivor Guilt in the Anthropocene. *Lapham's Quarterly.* URL: http://www.laphamsquarterly.org/disaster/human-error.

38 For a superb overview of the sixth mass extinction, see Sutter, John. 2016. Vanishing: the extinction crisis is far worse than you think. CNN. URL: http://www.cnn.com/interactive/2016/12/specials/vanishing/.

39 Barnosky, Anthony et al. 2012. Approaching a State Shift in Earth's Biosphere. *Nature.* 468: 52–58.

40 Personal communication.

41 See Richardson, Valerie. 2016. California Senate Sidelines Bill to Prosecute Climate Change Skeptics. *Washington Times.* URL: http://www.washingtontimes.com/news/2016/jun/2/calif-bill-prosecutes-climate-change-skeptics/.

42 Pinker, Steven. 2011. *The Better Angels of Our Nature: Why Violence Has Declined.* New York, NY: Penguin Books.

43 Ibid.

44 Incidentally, the futurist Jim Dator argues that democracy itself can lead to lower levels of concern for the future: "While I remain a tireless worker and advocate for more democracy, I have become increasingly aware that simply 'more democracy' alone, without other changes also accompanying it, does not solve all of the problems which concern me. While 'more democracy' means that disenfranchised or marginal people in the present might be able to have their needs properly attended to, 'more democracy,' even 'more direct democracy' as presently proposed, does

not also necessarily result in public thought and action more clearly oriented to the needs of future generations, I have observed. Indeed, more of what is often currently called 'democracy' seems to result in less future-orientation. Thus 'more democracy plus something else' is needed, I have concluded." Dator, Jim. 1999. Future Generatinos: They Are Our Conscience. In Tae-Chang Kim and Jim Dator (editors), *Co-Creating a Public Philosophy for Future Generations*. New York, NY: Praeger.

45 See Grossman, Daniel. 2016. High CO2 Levels Inside & Out: Double Whammy? *Yale Climate Connections*. URL: http://www.yaleclimateconnections.org/2016/07/indoor-co2-dumb-and-dumber/.

46 Ibid.

47 Ibid. One should also note that as CO2 concentrations rise outside, it is common for indoor settings to reach CO2 levels many times higher. This has *direct implications for intellectual work conducted in the closed spaces of modern buildings.*

48 This section is indebted to Ord, Toby, Rafaela Hillerbrand, and Anders Sandberg. 2010. Probing the Improbable: Methodological Challenges for Risks with Low Probabilities and High Stakes. *Journal of Risk Research*. 13: 191–205.

49 CERN. The safety of the LHC. Accessed on December 14, 2016. URL: https://press.cern/backgrounders/safety-lhc.

50 Ibid.

51 See Ord, Toby, Rafaela Hillerbrand, and Anders Sandberg. 2010. Probing the Improbable: Methodological Challenges for Risks with Low Probabilities and High Stakes. *Journal of Risk Research*. 13: 191–205.

52 Hut, Piet, and Martin Rees. 1983. How stable is our vacuum? *Nature*. 302: 508–509.

53 Tegmark, Max, and Nick Bostrom. 2005. Astrophysics: Is a Doomsday Catastrophe likely? *Nature*. 438: 754.

54 Ord, Toby, Rafaela Hillerbrand, and Anders Sandberg. 2010. Probing the Improbable: Methodological Challenges for Risks with Low Probabilities and High Stakes. *Journal of Risk Research*. 13: 191–205.

55 This concept is discussed more in subsection 4.3.1, point (vi).

56 See Hand, Eric. 2016. Could Bright, Foamy Wakes from Ocean Ships Combat Global Warming? *Science*. URL: http://www.sciencemag.org/news/2016/01/could-bright-foamy-wakes-ocean-ships-combat-global-warming.

57 Falk, Dan. 2017. Can Hacking the Planet Stop Runaway Climate Change? NBC. URL: http://www.nbcnews.com/mach/environment/ can-hacking-planet-stop-runaway-climate-change-n752221.

58 See Torres, Phil. 2017. The Only Way to Prevent Another Nuclear Strike Is to Get Rid of All the Nukes. *Motherboard*. URL: https://motherboard. vice.com/en_us/article/doomsday-clock-interview-lawrence-krauss.

59 Baum, Seth, Timothy Maher, and Jacob Haqq-Misra. 2013. Double Catastrophe: Intermittent Stratospheric Geoengineering Induced By Societal Collapse. *Environment, Systems and Decisions*. 33(1): 168–180.

60 Global Challenges Foundation. 2016. Global Catastrophic Risks 2016. URL: http://globalprioritiesproject.org/wp-content/uploads/2016/04/ Global-Catastrophic-Risk-Annual-Report-2016-FINAL.pdf.

61 Bostrom, Nick, Thomas Douglas, and Anders Sandberg. 2016. The Unilateralist's Curse and the Case for a Principle of Conformity. *Social Epistemology*. 30(4): 350–371.

62 Ibid.

63 Ibid.

64 Ibid.

Chapter 4

1 See, e.g., Shklovskii, I.S., and Carl Sagan. 1966. *Intelligent Life in the Universe*. New York, NY: Emerson-Adams Press.

2 Or, as Martin Rees puts it using a different metaphor, "Can the global village cope with its village idiots—especially when even one could be too many?" See Rees, Martin. 2008. Foreword. In Nick Bostrom and Milan Ćirković (editors), *Global Catastrophic Risks*. Oxford: Oxford University Press. See the beginning of section 4.3.3.

3 Gary Ackerman refers to a similar idea—I discovered only after publishing on the topic—in his PhD diss., calling it the "technology-organization dyad." See Ackerman, Gary. 2014. *"More Bang for the Buck": Examining the Determinants of Terrorist Adoption of New Weapons Technologies*. War Studies, King's College London. Also: in a book chapter coauthored with William Potter, Ackerman makes a plea to expand the scope of consideration to include agential motivations. They write, in the context of nuclear terrorism, that "an intentional risk such as nuclear terrorism can only result from the conscious decision of a human actor, yet the integral motivational component of such threats has often

been overshadowed by assessments of terrorist capabilities and weapon consequences." Ackerman, Gary, and William Potter. 2008. Catastrophic Nuclear Terrorism: A Preventable Peril. In Nick Bostrom and Milan Ćirković (editors), *Global Catastrophic Risks*. Oxford: Oxford University Press. Similarly, Ali Nouri and Christopher Chyba write that "to the extent that the biological security threat emanates from terrorist groups or irresponsible nations, a similar sophistication with respect to motives and behaviour must be brought to bear." Nouri, Ali, and Christopher Chyba. 2008. Biotechnology and Biosecurity. In Nick Bostrom and Milan Ćirković (editors), *Global Catastrophic Risks*. Oxford: Oxford University Press.

4   See Torres, Phil. 2016. Agential Risks: A Comprehensive Introduction. *Journal of Evolution and Technology*. 26(2): 31–47.

5   There are two possible conceptions of the word "coupled," depending on whether one focuses on the *afferent* (incoming information) or the *efferent* (outgoing information). The topic in this book is the latter, namely, how agents, whether biological or artificial in constitution, can use technologies as *extensions of themselves* to manipulate and rearrange the physical world. With respect to the former, though, the philosopher Don Ihde identifies two kinds of phenomenological relations that agents coupled to tools can have, namely, *embodiment* and *hermeneutic* relations. The first occurs when "technology is taken as the very medium of subjective perceptual experience of the world, thus transforming the subject's perceptual and bodily sense." Eyeglasses are an example of this, since they merge into our "phenomenological selves" as a quasi-transparent techno-modification of our visual systems. Or, consider what the philosopher Daniel Dennett writes about the phenomenology of driving, where the automobile is a kind of exoskeleton extension of our phenotypes: "Think of how you can feel the slipperiness of an oil spot on the highway under the wheels of your car as you turn a corner. The phenomenological focal point of contact is the point where the rubber meets the road, not any point on your innervated body, seated, clothed, on the car seat, or on your gloved hands on the steering wheel." (Dennett, Daniel. 1991. *Consciousness Explained*. New York, NY: Little Brown.) The second type of relation involves the technology functioning "as an immediate referent to something beyond itself." In this case, the technology itself is opaque, yet it enables us to "see" something beyond itself, as in the case of, say, a thermometer. For more on these and other phenomenological issues (including some that don't involve agent-tool couplings), see Introna, Lucas. 2011. Phenomenological Approaches to Ethics and Information Technology. *The Stanford Encyclo-*

pedia of Philosophy. URL: www.plato.stanford.edu/archives/sum2011/ entries/ethics-it-phenomenology, and Ihde, Don. 1990. *Technology and the Lifeworld: From garden to earth*. Bloomington and Indianapolis: Indiana University Press.

6   See Cirincione, Joseph. 2016. What Should the World Do with Its Nuclear Weapons? *Atlantic*. URL: https://www.theatlantic.com/international/archive/2016/04/global-nuclear-proliferation/478854/.

7   This phrase is borrowed from Torres, Phil. 2016. *The End: What Science and Religion Tell Us about the Apocalypse*. Charlottesville, VA: Pitchstone Publishing.

8   Roser, Max. 2016. Technological Progress. OurWorldInData.org. URL: https://ourworldindata.org/technological-progress/.

9   See Kurzweil, Ray. 2005. *The Singularity Is Near: When Humans Transcend Biology*. New York, NY: Penguin Books.

10   Hayden, Erika. 2014. Technology: The $1,000 Genome. *Nature*. URL: http://www.nature.com/news/technology-the-1-000-genome-1.14901.

11   Special Report. 2006. Life 2.0. *Economist*. URL: http://www.economist.com/node/7854314. Specifically, the Carlson curve describes "the rate of DNA sequencing or cost per sequenced base as a function of time" (to quote Wikipedia).

12   Roco, Mihail. 2011. The Long View of Nanotechnology Development: The National Nanotechnology Initiative at 10 Years. *Journal of Nanoparticle Research*. 13(2): 427–45. To be clear, the sort of nanotechnology discussed in subsection 4.2.2 is *molecular* nanotechnology, whereas these trends apply to nanotechnology in the sense of "nanoscale technologies" that are not necessarily atomically precise. Nonetheless, this data gestures at a general trend of exponential development. For more on nanotechnology, see also Allhoff, Fritz, Patrick Lin, and Daniel Moore. 2010. *What Is Nanotechnology and Why Does It Matter? From Science to Ethics*. New York, NY: Wiley-Blackwell.

13   See Kurzweil, Ray. 2005. *The Singularity Is Near: When Humans Transcend Biology*. New York, NY: Penguin Books.

14   Some of this material is borrowed from Torres, Phil. 2016. *The End: What Science and Religion Tell Us about the Apocalypse*. Charlottesville, VA: Pitchstone Publishing.

15   See, e.g., Traynor, Ian. 2008. Nuclear Bomb Blueprints for Sale on World Black Market, Experts Fear. *Guardian*. URL: https://www.theguardian.com/world/2008/may/31/nuclear.internationalcrime.

16 See Bostrom, Nick. 2011. Information Hazards: A Typology of Potential Harms from Knowledge. *Review of Contemporary Philosophy*. 10:44–79.

17 Indeed, one must balance the advantages and hazards of disseminating potentially dangerous information, ultimately deciding which is best through cost-benefit analyses.

18 See Mukunda, Gautam et al. 2009. What Rough Beast? Synthetic Biology, Uncertainty, and the Future of Biosecurity. *Politics and the Life Sciences*. 28(2).

19 Ibid.

20 Torres, Phil. 2016. Agential Risks: A New Direction for Existential Risk Scholarship. XRI Technical Report. URL: http://media.wix.com/ugd/d9aaad_a2bb594c8ead4ce5bbfc8e8a54534b50.pdf.

21 As the geneticist George Church writes, "In terms of threats, let's put aside nanotech and robots for the moment, there are two things that are more realistic threats, which are computer viruses and bio-terrorism. Those already exist. Nanotech and robots don't exist, unless you call a computer a robot—they're not developed enough to be a threat." Church, George. 2006. Constructive Biology. Edge.org. URL: https://www.edge.org/conversation/george_church-constructive-biology.

22 See, e.g., CTBTO. 2017. 1 November 1952—Ivy Mike. URL: https://www.ctbto.org/specials/testing-times/1-november-1952-ivy-mike.

23 Quoted from Torres, Phil. 2016. *The End: What Science and Religion Tell Us about the Apocalypse*. Charlottesville, VA: Pitchstone Publishing.

24 Cirincione, Joseph. 2013. *Nuclear Nightmares: Securing the World Before It Is Too Late*. New York, NY: Columbia University Press.

25 Ibid.

26 This threat could potentially grow in the future due to new technologies that facilitate the enrichment of uranium, such as SILEX, or the "Separation of Isotopes by Laser Excitation." As the physicist Ryan Snyder writes, "Third generation laser uranium enrichment technology, with the SILEX process being a likely example, may create new proliferation risks. . . . These conclusions suggest that third generation laser enrichment provides a new technological pathway to weapon-grade uranium and nuclear weapon development, with the acquisition of a usable laser being the main technological hurdle." Snyder, Ryan. 2016. A Proliferation Assessment of Third Generation Laser Uranium Enrichment Technology. *Science and Global Security*. 24(2): 68–91. Yet we should also note that, in terms of accessibility, "a U.S. government sponsored experi-

ment in the 1960s suggests that several physics graduates without prior experience with nuclear weapons and with access to only unclassified information could design a workable implosion type bomb." Ackerman, Gary, and William Potter. 2008. Catastrophic Nuclear Terrorism: A Preventable Peril. In Nick Bostrom and Milan Ćirković (editors), *Global Catastrophic Risks*. Oxford: Oxford University Press.

27 ISIS. 2015. The Perfect Storm. *Dabiq*. 9. URL: https://azelin.files. wordpress.com/2015/05/the-islamic-state-e2809cdc481biq-magazine-9e280b3.pdf.

28 Ibid.

29 Consider what scholars refer to as the "social amplification" of tragedies. To quote Cass Sunstein at length: "the loss of 200 million people may be more than 1,000 times worse than the loss of 2,000 people. Pause over the real-world meaning of a loss of 200 million people in the United States. The nation would find it extremely hard to recover. Private and public institutions would be damaged for a long time, perhaps forever. What kind of government would emerge? What would its economy look like? Future generations would inevitably suffer. The effect of a catastrophe greatly outruns a simple multiplication of a certain number of lives lost. The overall 'cost' of losing two-thirds of the American population is far more than 100,000 times the cost of losing 2,000 people. The same point holds when the numbers are smaller. Following the collapse of a dam that left 120 people dead and 4,000 homeless in Buffalo Creek, Virginia, psychiatric researchers continued to find significant psychological and sociological changes two years after the disaster occurred. Survivors still suffered a loss of direction and energy, along with other disabling character changes. One evaluator attributed this 'Buffalo Creek Syndrome' specifically to 'the loss of traditional bonds of kinship and neighborliness.' . . . Genuine catastrophes, involving tens of thousands or millions of deaths, would magnify that loss to an unimaginable degree." See Sunstein, Cass. 2007. *Worst-Case Scenarios*. Cambridge, MA: Harvard University Press. (Some paragraph breaks were deleted.)

30 Meade, Charles, and Roger Molander. 2006. Considering the Effects of a Catastrophic Terrorist Attack. RAND Corporation Technical Report. URL: http://www.rand.org/content/dam/rand/pubs/technical_reports/2006/RAND_TR391.pdf?. Note that parts of this section are excerpted from Torres, Phil. 2016. Apocalypse. . . When? It Matters Which Trend Lines One Follows: Why Terrorism *Is* an Existential Threat. *Free Inquiry*. URL: https://goo.gl/WKYo4h.

31 Ackerman, Gary, and William Potter. 2008. Catastrophic Nuclear Ter-

rorism: A Preventable Peril. In Nick Bostrom and Milan Ćirković (editors), *Global Catastrophic Risks*. Oxford: Oxford University Press.

32   Rezaei, Farhad. 2016. Shopping for Armageddon: Islamist Groups and Nuclear Terror. *Middle East Policy*. 13(3). URL: http://www.mepc.org/journal/middle-east-policy-archives/shopping-armageddon-islamist-groups-and-nuclear-terror?print.

33   Quoted in Cirincione, Joseph. 2013. *Nuclear Nightmares: Securing the World Before It Is Too Late*. New York, NY: Columbia University Press.

34   Ibid.

35   Ibid.

36   Allison, Graham. 2010. Nuclear Disorder: Surveying Atomic Threats. *Foreign Affairs*. 89(1): 74–85.

37   Quoted in ibid.

38   See Torres, Phil. 2016. *The End: What Science and Religion Tell Us about the Apocalypse*. Charlottesville, VA: Pitchstone Publishing; and Hamilton, Tyler. 2007. Capturing Carbon with Enzymes. *MIT Technology Review*. URL: https://www.technologyreview.com/s/407350/capturing-carbon-with-enzymes/.

39   See Horvath, Philippe, and Rodolphe Barrangou. 2010. CRISPR/Cas, the Immune System of Bacteria and Archaea. *Science*. URL: http://science.sciencemag.org/content/327/5962/167.full

40   Quoted in Saey, Tina Hesman. 2015. Gene Drives Spread Their Wings. *Science News*. URL: https://www.sciencenews.org/article/gene-drives-spread-their-wings.

41   Pollack, Andrew. 2002. Traces of Terror: The Science; Scientists Create a Live Polio Virus. *New York Times*. URL: http://www.nytimes.com/2002/07/12/us/traces-of-terror-the-science-scientists-create-a-live-polio-virus.html.

42   Tucker, Jonathan. 2011. Could Terrorists Exploit Synthetic Biology? *The New Atlantis*. 31: 69–81.

43   Van Aken, Jan. 2007. Ethics of Reconstructing Spanish Flu: Is It Wise to Resurrect a Deadly Virus? *Heredity*. 98: 1–2.

44   See Furmanski, Martin. 2014. Threatened Pandemics and Laboratory Escapes: Self-fulfilling Prophecies. *Bulletin of the Atomic Scientists*. URL: http://thebulletin.org/threatened-pandemics-and-laboratory-escapes-self-fulfilling-prophecies7016.

45   See Gholipour, Bahar. 2013. 2009 Swine-Flu Death Toll 10 Times Higher

Than Thought. *Live Science*. URL: http://www.livescience.com/41539-2009-swine-flu-death-toll-higher.html.

46   Nouri, Ali, and Christopher Chyba. 2008. Biotechnology and Biosecurity. In Nick Bostrom and Milan Ćirković (editors), *Global Catastrophic Risks*. Oxford: Oxford University Press. Note that this claim is outdated. A recent study finds that the bacterium *Vibrio natriengens* can replicate twice as fast. See Lee, Henry et al. 2016. Vibrio Natriengens, a New Genomic Powerhouse. *bioRxiv*. URL: http://biorxiv.org/content/early/2016/06/12/058487.

47   See Afshinnekoo, Ebrahim et al. 2015. Geospatial Resolution of Human and Bacterial Diversity with City-Scale Metagenomics. *Cell Systems*. 1: 1–15. I should be careful not to overstate the situation of anthrax in soil: one would also need to isolate the sample and determine if the strain is pathogenic to humans.

48   Global Challenges Foundation. 2016. Global Catastrophic Risks 2016. URL: http://globalprioritiesproject.org/wp-content/uploads/2016/04/Global-Catastrophic-Risk-Annual-Report-2016-FINAL.pdf.

49   Ibid.

50   See Global Challenges Foundation. 2015. 12 Risks That Threaten Human Civilisation. URL: https://api.globalchallenges.org/static/wp-content/uploads/12-Risks-with-infinite-impact.pdf.

51   See Allhoff, Fritz, Patrick Lin, and Daniel Moore. 2010. *What is Nanotechnology and Why Does it Matter? From Science to Ethics*. New York, NY: Wiley-Blackwell.

52   See Drexler, Eric. 1992. *Nanosystems: Molecular Machinery, Manufacturing, and Computation*. New York, NY: John Wiley & Sons; Drexler, Eric. 1986. *Engines of Creation: The Coming Era of Nanotechnology*. New York, NY: Anchor Books; and Drexler, Eric. 2013. *Radical Abundance: How a Revolution in Nanotechnology Will Change Civilization*. New York, NY: PublicAffairs.

53   This makes nanofactories qualitatively different from the sort of 3-D printers around today. For example, these employ "additive manufacturing" techniques, whereby melted plastic (or metal) is added one layer at a time. But a nanofactory could, by repositioning atoms, produce products made of virtually any material, since all macroscopic properties are reducible without remainder to the 3-D arrangement of their molecular and atomic constituents.

54   Phoenix, Chris, and Mike Treder. 2008. Nanotechnology as Global Catastrophic Risk. In Nick Bostrom and Milan Ćirković (editors), *Global*

Catastrophic Risks. Oxford: Oxford University Press.

55 Ibid.

56 See Drexler, Eric. 2013. *Radical Abundance: How a Revolution in Nano-technology Will Change Civilization*. New York, NY: PublicAffairs; Kurzweil, Ray. 2005. *The Singularity Is Near: When Humans Transcend Biology*. New York, NY: Penguin Books; and Diamandis, Peter. 2012. *Abundance: The Future Is Better Than You Think*. New York, NY: Free Press.

57 See Torres, Phil. 2016. *The End: What Science and Religion Tell Us about the Apocalypse*. Charlottesville, VA: Pitchstone Publishing, 79.

58 Wittes, Benjamin, and Gabriella Blum. 2015. *The Future of Violence: Robots and Germs, Hackers and Drones—Confronting a New Age of Threat*. New York, NY: Basic Books.

59 See Hobbes, Thomas. 1651 (1985). *Leviathan*. New York, NY: Penguin Books, 186.

60 For more on this issue, see Pinker, Steven. 2011. *The Better Angels of Our Nature: Why Violence Has Declined*. New York, NY: Penguin Books.

61 See also Wittes, Benjamin, and Gabriella Blum. 2015. *The Future of Violence: Robots and Germs, Hackers and Drones—Confronting a New Age of Threat*. New York, NY: Basic Books.

62 Pinker, Steven. 2011. *The Better Angels of Our Nature: Why Violence Has Declined*. New York, NY: Penguin Books.

63 Ibid.

64 Ibid.

65 Phoenix, Chris, and Mike Treder. 2008. Nanotechnology as Global Catastrophic Risk. In Nick Bostrom and Milan Ćirković (editors), *Global Catastrophic Risks*. Oxford: Oxford University Press.

66 To be clear, nanofactories aren't *themselves* a WTD. But they could nonetheless be employed for world-destroying purposes.

67 Feynman, Richard. 1959. Plenty of Room at the Bottom. URL: https://www.pa.msu.edu/~yang/RFeynman_plentySpace.pdf.

68 Alternatively, there could be "oceanic goo" that feeds on marine organisms. Indeed, Robert Freitas differentiates between "grey goo" (land-based replicators), "grey dust" (airborne replicators), "grey lichen" (chemolithotropic replicators), and "grey plankton" (ocean-borne replicators). See Freitas, Robert. 2000. Some Limits to Global Ecophagy by Biovorous Nanoreplicators, with Public Policy Recommendations. Foresight Institute. URL: http://www.rfreitas.com/Nano/Ecophagy.htm.

Note further that researchers initially thought that molecular manufacturing would require self-replication, but subsequent work has shown this not to be the case. Thus, Drexler argues that his *original* grey goo worries are now "obsolete." In his words, "The popular version of the grey-goo idea seems to be that nanotechnology is dangerous because it means building tiny self-replicating robots that could *accidentally* run away, multiply and eat the world. But there's no need to build anything remotely resembling a runaway replicator, which would be a pointless and difficult engineering task" (italics added). Quoted in In Depth. 2004. Drexler Dubs "Grey Goo" Fears Obsolete. Nanotechweb.org. URL: http://nanotechweb.org/cws/article/indepth/19648. Nonetheless, as the now-defunct Center for Responsible Nanotechnology writes, "goo type systems do not appear to be ruled out by the laws of physics, and we cannot ignore the possibility that the five stated requirements [that are necessary for nanobotic self-replication, namely, *mobility*, a *shell*, *control*, *metabolism*, and *fabrication*] could be combined *deliberately* at some point, in a device small enough that cleanup would be costly and difficult" (italics added). At the extreme, such an intentional attack "could make the earth largely uninhabitable," as Robert Freitas and Ralph Merkle put it. Freitas, Robert, and Ralph Merkle. 2004. *Kinematic Self-Replicating Machines*. Georgetown, TX: Landes Bioscience.

69   Kurzweil, Ray. 2005. *The Singularity Is Near: When Humans Transcend Biology*. New York, NY: Penguin Books.

70   For a comprehensive technical overview of the topic, see Freitas, Robert, and Ralph Merkle. 2004. *Kinematic Self-Replicating Machines*. Georgetown, TX: Landes Bioscience.

71   Probably *after* synthetic biology reaches maturity but *before* superintelligence. For example, Ray Kurzweil writes that "the window of malicious opportunity for bioengineered viruses, existential or otherwise, will close in the 2020s when we have fully effective antiviral technologies based on nanobots. However, because nanotechnology will be thousands of times stronger, faster, and more intelligent than biological entities, self-replicating nanobots will present a greater risk and yet another existential risk. The window for malevolent nanobots will ultimately be closed by strong artificial intelligence, but, not surprisingly, 'unfriendly' AI will itself present an even more compelling existential risk." Kurzweil, Ray. 2006. *The Singularity Is Near: When Humans Transcend Biology*. New York, NY: Viking. As discussed in section 6.4, though, this may not be the most optimal order of arrival.

72   Quoted in Topol, Sarah. 2016. Attack of the Killer Robots. *Buzzfeed*.

URL: https://www.buzzfeed.com/sarahatopol/how-to-save-mankind-from-the-new-breed-of-killer-robots?utm_term=.ecQOyd1WLm#.klbdvNgx26.

73 Quoted in ibid.

74 Asaro, Peter. 2013. On Banning Autonomous Weapon Systems: Human Rights, Automation, and the Dehumanization of Lethal Decision-Making. *New Technologies and Warfare.* 94(886): 687–709.

75 To be clear, Conn is summarizing a view put forward by the computer scientist Oren Etzioni during a White House conference about AI in 2016. See Conn, Ariel. 2016. The White House Considers the Future of AI. Future of Life Institute. URL: http://futureoflife.org/2016/06/15/white-house-future-ai-1/.

76 Yudkowsky, Eliezer. 2008. Cognitive Biases Potentially Affecting Judgement of Global Risks. In Nick Bostrom and Milan Ćirković (editors), *Global Catastrophic Risks.* Oxford: Oxford University Press.

77 Note that "Islamist" has a very *specific* definition, namely, one who sees the Quran as a political model upon which the governmental system should be founded. In other words, Islamists reject the separation of mosque and state. Perhaps only 10 percent of the global Muslim population adhere to Islamist beliefs, and an even smaller percentage sees violence as a legitimate tool for establishing a theocracy. For more, see Harris, Sam, and Maajid Nawaz. 2015. *Islam and the Future of Tolerance: A Dialogue.* Cambridge, MA: Harvard University Press.

78 UNICEF. 1999. Results of the 1999 Iraq Child and Maternal Mortality Surveys. URL: http://reliefweb.int/report/iraq/results-1999-iraq-child-and-maternal-mortality-surveys.

79 I do not want to claim that the following enumeration is exhaustive. Research on this topic is still inchoate, and as such the present book may be missing additional types—or important subtypes—of agential risks. For example, I have spoken with some people—essentially transhumanists—who have said that they would rather humanity be annihilated in a flash than for us to never reach a posthuman state. Along these lines, one might be tempted to push an "extinction button" if it appears, to them, highly likely that humanity is about to slip into a state of permanent stagnation. The motivation here may even be in some sense ethical. Since this book is intended to be, in large part, an intellectual springboard for future scholarship, I strongly encourage readers to take the following analyses a step further!

80 Institute for Economics and Peace. 2016. Global Terrorism Index. URL:

http://economicsandpeace.org/wp-content/uploads/2016/11/Global-Terrorism-Index-2016.2.pdf. For a helpful discussion of the topic, see Arnett, George. 2014. Religious Extremism Main Cause of Terrorism, According to Report. *Guardian*. URL: https://www.theguardian.com/news/datablog/2014/nov/18/religious-extremism-main-cause-of-terrorism-according-to-report.

81 For example, see Rausch, Cassandra Christina. 2015. Fundamentalism and Terrorism. *Journal of Terrorism Research*. 6(2). URL: http://jtr.st-andrews.ac.uk/articles/10.15664/jtr.1153/; and Torres, Phil. 2016. Apocalypse. . . When? It Matters Which Trend Lines One Follows: Why Terrorism *Is* an Existential Threat. *Free Inquiry*. URL: https://goo.gl/WKYo4h.

82 Institute for Economics and Peace. 2016. Global Terrorism Index. URL: http://economicsandpeace.org/wp-content/uploads/2016/11/Global-Terrorism-Index-2016.2.pdf.

83 Or, as I have written elsewhere, "For them, what matters isn't this life, but the afterlife; the ultimate goal isn't worldly, but otherworldly. These unique features make religious terrorism especially dangerous." Torres, Phil. 2016. Agential Risks: A Comprehensive Introduction. *Journal of Evolution and Technology*. 26(2): 31–47. Similarly, the terrorism scholars Jessica Stern and JM Berger observe that, "violent apocalyptic groups are not inhibited by the possibility of offending their political constituents because they see themselves as participating in the ultimate battle." Consequently, they are "*the most likely terrorist groups to engage in acts of barbarism, and to attempt to use rudimentary weapons of mass destruction*" (italics added). See Stern, Jessica, and J.M. Berger. 2015. *ISIS: The State of Terror*. New York, NY: HarperCollins Publishers.

84 See Landes, Richard. 2011. *Heaven on Earth: The Varieties of the Millennial Experience*. Oxford: Oxford University Press.

85 The *most deadliest war*, namely, World War II, also involved a version of active apocalypticism, as discussed in section 1.7.

86 That is, according to the "two-seed" strain of Christian Identity.

87 See Flannery, Frances. 2016. *Understanding Apocalyptic Terrorism: Countering the Radical Mindset*. New York, NY: Routledge.

88 Ibid.

89 See Wood, Graeme. 2014. The Three Types of People Who Fight for ISIS. *New Republic*. URL: https://newrepublic.com/article/119395/isiss-three-types-fighters.

90 The "hadith" are a record of the sayings and actions of the prophet Muhammad. For more, see Torres, Phil. 2016. *The End: What Science and Religion Tell Us about the Apocalypse*. Charlottesville, VA: Pitchstone Publishing.

91 See McCants, William. 2016. *The ISIS Apocalypse: The History, Strategy, and Doomsday Vision of the Islamic State*. New York, NY: St. Martin's Press.

92 Specifically, the hadith states: "'There will be Prophethood for as long as Allah wills it to be, then He will remove it when He wills, then there will be Khilafah [i.e., Caliphate] on the Prophetic method and it will be for as long as Allah wills, then He will remove it when He wills, then there will be biting Kingship for as long as Allah Wills, then He will remove it when He wills, then there will be oppressive kingship for as long as Allah wills, then he will remove it when He wills, and then there will be Khilafah upon the Prophetic method' and then he remained silent." Editorial. 2014. Daily Hadith. URL: http://www.khilafah.com/daily-hadith-256/.

93 In fact, Daesh's primary propaganda magazine is called *Dabiq*. And their Turkish language propaganda magazine is called *Konstantiniyye*, which translates as Constantinople (see section 1.7).

94 BBC Monitoring. 2016. Dabiq: Why Is Syrian Town So Important for IS? BBC. URL: http://www.bbc.com/news/world-middle-east-30083303.

95 For example, Christian dispensationalists believe that Christians will be raptured before Armageddon, some Sunnis anticipate that one-third of the Muslim army in Dabiq will survive the Grand Battle, and, as just stated, Aum held that the coming total war would leave their group intact. See Torres, Phil. 2016. Agential Risks: A Comprehensive Introduction. *Journal of Evolution and Technology*. 26(2): 31–47.

96 It is worth registering, if only in an endnote, that history reveals many doomsday cults that turned their violence *inward* on themselves, rather than *outward* on society. Heaven's Gate provides an example: its members ingested phenobarbital mixed with applesauce and downed with vodka in order to board a UFO that they believed was trailing the Hale-Bopp comet. The result was a mass suicide event that left 39 people dead. Although this group held a kind of *active apocalyptic* ideology, they were motivated by neither homicidal nor omnicidal impulses. Thus Heaven's Gate, presumably, would not have pushed a doomsday button if it were available. See Torres, Phil. 2016. Apocalypse Soon? How Emerging Technologies, Population Growth, and Global Warming Will Fuel Apocalyptic Terrorism in the Future. *Skeptic*. URL: https://goo.gl/

rIwJkm; and Torres, Phil. 2016. Apocalypse. . . When? It Matters Which Trend Lines One Follows: Why Terrorism *Is* an Existential Threat. *Free Inquiry*. URL: https://goo.gl/WKYo4h.

97 Thus, the term "misguided" is specifically relativized to the normative framework in which we are working: achieving the particular goal of technological maturity, as previously defined.

98 Benatar, David. 2006. *Better Never to Have Been: The Harm of Coming into Existence*. Oxford: Oxford University Press.

99 This risk type has been noted in the existential risk literature at least since Leslie, John. 1996. *The End of the World: The Science and Ethics of Human Extinction*. New York, NY: Routledge.

100 See Chao, Roger. 2012. Negative Average Preference Utilitarianism. *Journal of Philosophy of Life*. 2(1): 55–66.

101 There are, in fact, many other shades of NU too, such as "weak," "lexical threshold," "lexical," and "absolute" NU, the latter of which is equivalent to SNU, as used in this book. For details, see Ord, Toby. 2013. Why I'm Not a Negative Utilitarian. URL: http://www.amirrorclear.net/academic/ideas/negative-utilitarianism/. David Pearce also identifies "negative ideal preference utilitarianism" and "negative hedonistic utilitarianism," the latter of which has both "lexical" and "Buddhist" versions. See Pearce, David. Negative Utilitarianism FAQ. Accessed on May 3, 2017. URL: https://www.utilitarianism.com/nu/nufaq.html.

102 Pearce, David. Negative Utilitarianism. Accessed February 5, 2017. URL: https://www.hedweb.com/negutil.htm. One may be reminded here of Emmanuel Levinas, who once declared that the Holocaust is the "end of theodicy," writing: "Harsh reality (this sounds like a pleonasm!) . . .," as well as E.E. Cummings colorful flourish that life is "the strenuous briefness"—sentiments that echo Pearce's plaintive assertions. See Levinas, Emmanuel. 1969. *Totality and Infinity: An Essay on Exteriority*. Boston, MA: Kluwer Academic Publishers; and Cummings, E.E. Into the Strenuous Briefness. Accessed on April 18, 2017. URL: https://hellopoetry.com/poem/1606/into-the-strenuous-briefness/.

103 Smart, R.N. 1958. Negative Utilitarianism. *Mind*. 67(268): 542–543. Note that Smart's criticism specifically focuses on "the human race," rather than sentient life in general. David Pearce, a negative utilitarian *and* transhumanist, responds to the world-exploder idea goes as follows: "If the destruction could be accomplished painlessly, then a negative utilitarian is logically compelled to accept this consequence. . . . However, planning and implementing the extinction of all sentient life couldn't

be undertaken painlessly. Even contemplating such an enterprise would provoke distress. Thus a negative utilitarian is not compelled to argue for the apocalyptic solution. S/he may still privately believe that it would have been better if the world had never existed. This is a separate issue." Pearce, David. The Pinprick Argument. Accessed on April 16, 2017. URL: https://www.utilitarianism.com/pinprick-argument.html. See also Pearce, David. Negative Utilitarianism FAQ. Accessed on May 3, 2017. URL: https://www.utilitarianism.com/nu/nufaq.html for discussion of the world-exploder issue.

104 Ibid.

105 See Death of Alan Kurdi. Wikipedia. Accessed on March 25, 2017. URL: https://en.wikipedia.org/wiki/Death_of_Alan_Kurdi.

106 Although, for reasons not here explicated, Pearce is a passionate advocate of existential risk reduction.

107 Some italics were removed. See Pearce, David. Unsorted Postings. Accessed on April 16, 2017. URL: https://www.hedweb.com/social-media/pre2014.html.

108 Ibid.

109 Quoted in Langman, Peter. 2014. Influences on the Ideology of Eric Harris. URL: https://schoolshooters.info/sites/default/files/harris_influences_ideology_1.2.pdf.

110 Quoted in Langman, Peter. 2009. *Why Kids Kill: Inside the Minds of School Shooters*. New York, NY: Palgrave Macmillan.

111 Langman, Peter. 2014. Influences on the Ideology of Eric Harris. URL: https://schoolshooters.info/sites/default/files/harris_influences_ideology_1.2.pdf. Note that while Harris appears to have been a sadistic sociopath, Klebold very likely suffered from psychosis.

112 Quoted in ibid.

113 Quoted in ibid. Note that this section borrows heavily from Torres, Phil. 2016. Agential Risks: A Comprehensive Introduction. *Journal of Evolution and Technology*. 26(2): 31–47.

114 See Page, Jonathan, Jeffrey Daniels, and Steven Craig. 2015. *Violence in Schools*. New York, NY: Springer.

115 Others, such as Kip Kinkel and Andrew Golden, were outright bullies who "harassed their peers. They were the perpetrators, not the victims, of threats and insults. They were seen as scary and intimidating kids whom it was best to avoid." Langman, Peter. 2009. *Why Kids Kill: Inside*

the Minds of School Shooters. New York, NY: Palgrave Macmillan.

116 To be clear, violence isn't a symptom of schizophrenia, and indeed schizophrenic people are much more likely to hurt themselves than others. See Schizophrenia and Poverty, Crime and Violence. Accessed on April 13, 2017. URL: http://schizophrenia.com/poverty.htm#.

117 Dictionary.com. Conscience. Accessed on March 7, 2017. URL: http://www.dictionary.com/browse/conscience?r=75&src=ref&ch=dic.

118 Stout, Martha. 2005. *The Sociopath Next Door*. New York, NY: Broadway Books.

119 Ibid.

120 See Associated Press. 2004. Man Who Bulldozed through Colo. Town Is Dead. NBC News. URL: http://www.nbcnews.com/id/5139598/ns/us_news-crime_and_courts/t/man-who-bulldozed-through-colo-town-dead/#.WLx_XxIrI4o.

121 It is, as I have written elsewhere, "also worth pointing out that Heemeyer saw himself as God's servant. As he put it, 'God blessed me in advance for the task that I am about to undertake. It is my duty. God has asked me to do this. It's a cross that I am going to carry and I'm carrying it in God's name.'" This suggests that he might have suffered from psychotic delusions. Torres, Phil. 2016. Agential Risks: A Comprehensive Introduction. *Journal of Evolution and Technology*. 26(2): 31–47.

122 On the one hand, "early defenders of Heemeyer contended he made a point of not hurting anybody during his bulldozer rampage." On the other, he "fired 15 bullets from his .50-BMG rifle at power transformers and propane tanks." Thus, as the local sheriff's department noted, "Had these tanks ruptured and exploded, anyone within one-half mile of the explosion could have been endangered." *The Durango Telegraph*. 2004. Dozer Rampage Roots Run Deep. Accessed on May 2, 2017. URL: http://archives.durangotelegraph.com/04-06-24/mountain_exchange.htm.

123 Thus, if Heemeyer had been, say, a microbiologist rather than a welder, and if his grudge had been against society in general rather than his local town in particular, who knows how devastating the consequences could have been.

124 The category of idiosyncratic actors should also include individuals who commit "crimes of passion" and outright misanthropes who simply wish for humanity's demise. With respect to the latter, Immanuel Kant writes the following: "Misanthropy is hatred of mankind, and takes two forms: aversion from men, and enmity towards them. In the first case we are

afraid of men, regarding them as our enemies; but the second is when a man is himself an enemy to others. The aversive man shrinks from men out of temperament, he sees himself as no good to others, and thinks he is too unimportant for them; and since, for all that, he has a certain love of honour, he hides and runs away from people. The enemy of mankind shuns his fellows on principle, thinking himself too good for them. Misanthropy arises, partly from dislike, and partly from ill-will. The misanthrope from dislike thinks all men are bad; he fails to find in them what he was seeking; he does not hate them, and wishes some good to all, but simply does not like them. Such people are melancholy folk, who can form no conception of the human race. But the misanthrope from ill-will is he who does good to nobody, and pursues their harm instead." Kant, Immanuel. 1997. *Lectures on Ethics*. Cambridge: Cambridge University Press. Note that misanthropy also overlaps with the category of ecoterrorism and, to some extent, apocalyptic terrorism.

125 Brennan, Andrew, and Yeuk-Sze Lo. 2016. Environmental Ethics. *The Stanford Encyclopedia of Philosophy*. URL: https://plato.stanford.edu/entries/ethics-environmental/.

126 Indeed, this is the dubious step!

127 VHEMT. About the Movement. Accessed February 5, 2017. URL: http://www.vhemt.org/aboutvhemt.htm.

128 Note that the founder of VHEMT, Les U. Knight, is explicitly motivated by deep ecology philosophy.

129 Milbank, Dana. 1994. A Strange Finnish Thinker Posits War, Famine as Ultimate "Goods." *Asian Wall Street Journal*.

130 Linkola, Pentti. 2004. *Can Life Prevail?* Helsinki: Tammi Publishers. He adds that "emphasis on the inalienable right to life or foetuses, premature infants and the brain-dead has become a kind of collective mental illness."

131 Milbank, Dana. 1994. A Strange Finnish Thinker Posits War, Famine as Ultimate "Goods." *Asian Wall Street Journal*; and Linkola, Pentti. 2004. *Can Life Prevail?* Helsinki: Tammi Publishers, respectively.

132 Milbank, Dana. 1994. A Strange Finnish Thinker Posits War, Famine as Ultimate "Goods." *Asian Wall Street Journal*. Note that Linkola claims *not* to wish for *total* human extinction like VHEMT, but he does unequivocally advocate for the mass slaughter of huge numbers of humans around the world. For example, he writes in a book passage (unverified by the present author) that "we even have to be able to re-evaluate Fascism and recognize the service that philosophy made 30 years ago when

it freed the Earth from the weight of tens of millions of overeating Europeans, six million of them by an almost ideally painless, environment-preserving means." This is, indeed, consistent with his "eco-fascism" and "lifeboat" philosophy, borrowed and modified from Garrett Hardin, which goes as follows: "What to do, when a ship carrying a hundred passengers suddenly capsizes and only one lifeboat, with room for only ten people, has been launched? When the lifeboat is full, those who hate life will try to load it with more people and sink the lot. Those who love and respect life will take the ship's axe and sever the extra hands that cling to the sides of the boat." Quoted in Linkola, Pentti. 2004. *Can Life Prevail?* Helsinki: Tammi Publishers.

133 Milbank, Dana. 1994. A Strange Finnish Thinker Posits War, Famine as Ultimate "Goods." *Asian Wall Street Journal.* Italics added.

134 Flannery, Frances. 2016. *Understanding Apocalyptic Terrorism: Countering the Radical Mindset.* New York, NY: Routledge. See also Media. 2010. Save the Planet Protest: James J. Lee's Demands (TEXT). *Huffington Post.* URL: http://www.huffingtonpost.com/2010/09/01/save-the-planet-protest-j_n_702781.html. One may be reminded here of the anarchist Edward Abbey's startling assertion that "it will be objected that a constantly increasing population makes resistance and conservation a hopeless battle. This is true. Unless a way is found to stabilize the nation's population, the parks can not be saved. Or anything else worth a damn. Wilderness preservation, like a hundred other good causes, will be forgotten under the overwhelming pressure of a struggle for mere survival and sanity in a completely urbanized, completely industrialized, ever more crowded environment. For my own part *I would rather take my chances in a thermonuclear war than live in such a world.*" Abbey, Edward. 1988. *Desert Solitaire.* Tucson, AZ: University of Arizona Press. Italics added.

135 Gaia Liberation Front. Statement of Purpose (A Modest Proposal). Accessed February 5, 2017. URL: http://www.churchofeuthanasia.org/resources/glf/glfsop.html.

136 1989. Eco-Kamikazes Wanted. *Earth First! Journal.* Also quoted in Dye, LaVonne. 1993. The Marine Mammal Protection Act: Maintaining the Commitment to Marine Mammal Conservation. *Case Western Reserve Law Review.* 43(4): 1411–1448. Thanks to Gary Ackerman for apprising me of this article. It appears to be somewhat sardonic, at times, but also quite serious.

137 Note that such individuals may or may not be inspired by the deep ecology movement *per se.* There are also neo-Luddites who advocate for

much more limited, specific technological restrictions, such as Chellis Glendinning, whose primary complaints concern nuclear and chemical technologies, genetic engineering, television, electromagnetic artifacts, and computers. See Glendinning, Chellis. 1990. Notes toward a neo-Luddite Manifesto. URL: https://theanarchistlibrary.org/library/chellis-glendinning-notes-toward-a-neo-luddite-manifesto.

138 Mumford, Lewis. 1967. *The Myth of the Machine: Technics and Human Development*. San Diego, CA: Harcourt Brace Jovanovich. Note that, according to a commentator of Kaczynski, while "the Unabomber doesn't properly acknowledge deep ecology ideas," his "implied approval of 'the idea that wild nature is more important than human economic welfare' places him close to deep ecology, if only to a misanthropic, catastrophist variety of it." See Fulano, T. 1996. The Unabomber and the Future of Industrial Society. *Fifth Estate*. 348: 5–9. Kaczynski himself wrote in his manifesto that "since there are well-developed environmental and wilderness movements, we have written very little about environmental degradation or the destruction of wild nature, even though we consider these to be highly important." Kaczynski, Ted. 1996. Unabomber Manifesto. URL: https://partners.nytimes.com/library/national/unabom-manifesto-1.html.

139 For more, see Flannery, Frances. 2016. *Understanding Apocalyptic Terrorism: Countering the Radical Mindset*. New York, NY: Routledge.

140 Quoted in Long, Douglas. 2004. *Ecoterrorism*. New York, NY: Facts On File.

141 Kaczynski, Ted. 1996. Unabomber Manifesto. URL: https://partners.nytimes.com/library/national/unabom-manifesto-1.html. For a detailed exploration of Kaczynski's vision for the future, see Kaczynski, Ted. 2016. *Anti-Tech Revolution: Why and How*. Scottsdale, AZ: Fitch and Madison Publishers. Note also that Kaczynski talked about a "wilderness religion," writing that "there is a religious vacuum in our society that could perhaps be filled by a religion focused on nature in opposition to technology." Quoted in Flannery, Frances. 2016. *Understanding Apocalyptic Terrorism: Countering the Radical Mindset*. New York, NY: Routledge. Finally, Kaczynski was once diagnosed with schizophrenia by a court-appointed psychiatrist, although this diagnosis is contentious. The point is that the categories of agential risks here enumerated are, indeed, distinct but overlapping.

142 Beckhusen, Robert. 2013. In Manifesto, Mexican Eco-Terrorists Declare War on Nanotechnology. *Wired*. URL: https://www.wired.com/2013/03/mexican-ecoterrorism/; and Loyd, Marion, and Jeffrey Young. 2011.

Nanotechnologists Are Targets of Unabomber Copycat, Alarming Universities. *Chronicle of Higher Education.* URL: http://www.chronicle.com/article/Nanotechnologists-Are-Targets/128764/.

143 Individualidades Tendiendo a lo Salvaje. 2013. Reivindicación del envió de paquete bomba a investigador de la UNAM y asesinato de biotecnólogo en 2011 en México. URL: http://liberaciontotal.lahaine.org/?p=4945. The anarcho-primitivist John Zerzan confirms this, saying, "The ITS group is real slavish to Ted Kaczynski." Morin, Roc. 2014. The Anarcho-Primitivist Who Wants us all to Give Up Technology. *Vice.* URL: https://www.vice.com/en_us/article/john-zerzan-wants-us-to-give-up-all-of-our-technology.

144 A quick Google search reveals an unsettling number of websites and discussion board posts on which many anonymous individuals affirm that they would destroy humanity *if only* the means for doing this were available. The ideological reasons most often given, on my reading, are (a) misanthropic, (b) environmentalist, and (c) ethical (along the lines of SNU). For example, a Debate.org question asks, "If you could push a button and destroy all human life. [*sic*] Would you? All other life would survive as is; plus evidence of mankind too." *Seventy-nine percent* of respondents answered "Yes," whereas only 21 percent answered "no." This does not need to be a scientific poll to warrant serious concern. For example, one person writes, "My view is that Mankind is a plague. . . . I vote to destroy mankind and let nature start over." Someone else adds, "The human animal is the only evil animal in the animal kingdom. We destroy everything. . . . I email the president weekly and beg him to push the button and stop the madness already." Along these lines, two other respondents say, "Yes i [*sic*] would because animals would be much better off without us and they are innocent beings that dont [*sic*] deserve the misery we bring them" and "In the short time we've been on this planet, humans have already destroyed so much. We destroy ecosystems, and kill off entire species of animals. . . . The world would be better off without humans as a whole." And finally, someone exhorts, "Stop the pain. I think it's egiostic [*sic*] to not destroy the world. Of course there are people who have a nice life. But there are other people and animals who are suffering and some have horrifying pain. I just don't think it will ever get better. There are to many [*sic*] people and to many [*sic*] rotten evil creatures." So, perhaps there are more people than one might otherwise expect in the world who would only consider perpetrating an act of destruction if a doomsday button were presented to them. See Debate.org. Accessed on April 21, 2017. URL: http://www.debate.org/opinions/if-you-could-push-a-button-and-destroy-all-hu-

man-life-would-you-all-other-life-would-survive-as-is-plus-evidence-of-mankind-too.

145 See The Top Myths About Advanced AI. Future of Life Institute. Accessed on December 29, 2016. URL: http://futureoflife.org/background/aimyths/.

146 Kolodny, Niko, and John Brunero. 2016. Instrumental Rationality. *The Stanford Encyclopedia of Philosophy*. URL: https://plato.stanford.edu/entries/rationality-instrumental/.

147 Bostrom, Nick. 2012. The Superintelligent Will: Motivation and Instrumental Rationality in Advanced Artificial Agents. *Minds and Machines*. 22(2): 71–85.

148 To be clear, the suggestion here isn't that this is likely to happen but that it's possible, and the very possibility of high levels of intelligence combined with arbitrary final goals should be sufficient for concern.

149 Bostrom, Nick. 2014. *Superintelligence: Paths, Dangers, Strategies*. Oxford: Oxford University Press. See also Omohundro, Stephen. 2008. The Basic AI Drives. In Pei Wang, Ben Goertzel, and Stan Franklin (editors), *Artificial General Intelligence 2008: Proceedings of the First AGI Conference*. Amsterdam: IOS Press.

150 As Stuart Russell puts it, "A machine with a specific purpose has another property, one that we usually associate with living things: a wish to preserve its own existence. For the machine, this trait is not innate, nor is it something introduced by humans; it is a logical consequence of the simple fact that the machine cannot achieve its original purpose if it is dead. So if we send out a robot with the sole directive of fetching coffee, it will have a strong incentive to ensure success by disabling its own off switch or even exterminating anyone who might interfere with its mission. If we are not careful, then, we could face a kind of global chess match against very determined, superintelligent machines whose objectives conflict with our own, with the real world as the chessboard. The prospect of entering into and losing such a match should concentrate the minds of computer scientists. . . . Some argue that . . . we can just 'switch them off' as if superintelligent machines are too stupid to think of that possibility." See Russell, Stuart. 2016. Should We Fear Supersmart Robots? *Scientific American*. 314: 58–59.

151 See Bostrom, Nick. 2014. *Superintelligence: Paths, Dangers, Strategies*. Oxford: Oxford University Press.

152 Even if we were able to escape this ignominious end, the paper clip maximizer would likely dismantle the biosphere to harvest the atoms that *it*

contains, thereby destroying the natural environment upon which humanity depends for survival.

153 For a helpful overview of this issue, see Harris, Sam. 2016. Can We Build AI without Losing Control over It? TED. URL: https://www.ted.com/talks/sam_harris_can_we_build_ai_without_losing_control_over_it.

154 Yudkowsky, Eliezer. 2013. Intelligence Explosion Microeconomics. Machine Intelligence Research Institute. URL: https://intelligence.org/files/IEM.pdf.

155 Good, I.J. 1966. Speculations concerning the First Ultraintelligent Machine. *Advances in Computers*. 6: 31–88.

156 This is sometimes referred to as the "singularity," although this term has accrued a multiplicity of meanings over the years. For more on the singularity as an *intelligence explosion*, see Chalmers, David. 2010. The Singularity: A Philosophical Analysis. *Journal of Consciousness Studies*. 17(9–10): 7–65.

157 Those skeptical of (3) and (4) should heed what the Asilomar conference calls the "Capability Caution Principle." This states that "there being no consensus, we should avoid strong assumptions regarding upper limits on future AI capabilities." As Ariel Conn elaborates, "Some experts think human-level or even super-human AI could be developed in a matter of a couple decades, while some don't think anyone will ever accomplish this feat. The Capability Caution Principle argues that, until we have concrete evidence to confirm what an AI can someday achieve, it's safer to assume that there are no upper limits—that is, for now, anything is possible and we need to plan accordingly." See Conn, Ariel. 2017. How Smart Can AI Get? Future of Life Institute. URL: https://futureoflife.org/2017/02/17/capability-caution-principle/. Note also that some of this is excerpted from Torres, Phil. 2016. We're Not Ready for Superintelligence. *Motherboard*. URL: http://motherboard.vice.com/read/superintelligence-and-fear-mongering-should-we-really-be-worried.

158 That is to say, a superintelligence's values could drift over time. Indeed, human values have done precisely this, and most would agree that our present values, largely inherited from the Enlightenment, are far superior to those that motivated people during the medieval or Paleolithic eras. What must not drift is some sort of *meta-value* that constrains the space of all other motivating values held by the superintelligence. This could, potentially, ensure that these motivating values are never inconsistent with what we value. See also Yudkowsky, Eliezer. 2004. Coherent Extrapolated Volition. Machine Intelligence Research Institute. URL: https://intelligence.org/files/CEV.pdf.

159 Bostrom, Nick. 2014. *Superintelligence: Paths, Dangers, Strategies*. Oxford: Oxford University Press. Alternatively, we could make an AI "learn" our values; but this comes with pitfalls of its own.

160 Note that Yudkowsky distinguishes between *technical* and *philosophical* failures of AI. In his words, "Technical failure is when you try to build an AI and it doesn't work the way you think it does—you have failed to understand the true workings of your own code. Philosophical failure is trying to build the wrong thing, so that even if you succeeded you would still fail to help anyone or benefit humanity. Needless to say, the two failures are not mutually exclusive." See Yudkowsky, Eliezer. 2008. Artificial Intelligence as a Positive and Negative Factor in Global Risk. In Nick Bostrom and Milan Ćirković (editors), *Global Catastrophic Risks*. Oxford: Oxford University Press.

161 Readers may recognize the second number anagram.

162 For a related scenario, see Robert Nozick's "experience machine" thought experiment, which aims to undercut hedonism. Nozick, Robert. 1974. *Anarchy, State, and Utopia*. New York, NY: Basic Books. Thanks to Daniel Kokotajlo for bringing my attention to this scenario.

163 Not to mention that being brains in a vat would prevent humanity from reaching a posthuman state.

164 A related issue is *infrastructure profusion*. This is a malignant failure mode whereby "an agent transforms large parts of the reachable universe into infrastructure in the service of some goal, with the side effect of preventing the realization of humanity's axiological potential." It could entail an existential catastrophe *even if* the superintelligence's final goals *aren't* open-ended (as is the case with a paper clip maximizer). For example, "suppose that the goal is instead to make at least one million paperclips (meeting suitable design specifications) rather than to make as many as possible. One would like to think that an AI with such a goal would build one factory, use it to make a million paperclips, and then halt. Yet this may not be what would happen. Unless the AI's motivation system is of a special kind, or there are additional elements in its final goal that penalize strategies that have excessively wide-ranging impacts on the world, there is no reason for the AI to cease activity upon achieving its goal. On the contrary: if the AI is a sensible Bayesian agent, *it would never assign exactly zero probability to the hypothesis that it has not yet achieved its goal*—this, after all, being an empirical hypothesis against which the AI can have only uncertain perceptual evidence. The AI should therefore continue to make paperclips in order to reduce the (perhaps astronomically small) probability that it has somehow still

failed to make at least a million of them, all appearances notwithstanding. There is nothing to be lost by continuing paperclip production and there is always at least some microscopic probability increment of achieving its final goal to be gained." See Bostrom, Nick. 2014. *Superintelligence: Paths, Dangers, Strategies*. Oxford: Oxford University Press.

165 Again, the following paper is relevant to this issue: Yudkowsky, Eliezer. 2004. Coherent Extrapolated Volition. Machine Intelligence Research Institute. URL: https://intelligence.org/files/CEV.pdf.

166 See Bourget, David, and David Chalmers. 2014. What Do Philosophers Believe? *Philosophical Studies*. 170(3): 465–500.

167 See Torres, Phil. 2016. We're Not Ready for Superintelligence. *Motherboard*. URL: http://motherboard.vice.com/read/superintelligence-and-fear-mongering-should-we-really-be-worried.

168 McGinn, Colin. 1993. *Problems in Philosophy: The Limits of Inquiry*. New York, NY: Wiley-Blackwell.

169 Bostrom, Nick. 2014. *Superintelligence: Paths, Dangers, Strategies*. Oxford: Oxford University Press.

170 Ibid.

171 Russell, Stuart. 2016. Should We Fear Supersmart Robots? *Scientific American*. 314: 58–59. For an important recent article about "safely interruptible agents," see Orseau, Laurent, and Stuart Armstrong. 2016. Safely Interruptible Agents. 32nd Conference on Uncertainty in Artificial Intelligence. URL: https://intelligence.org/files/Interruptibility.pdf.

172 See Armstrong, Stuart, Anders Sandberg, and Nick Bostrom. 2012. Thinking Inside the Box: Controlling and Using an Oracle AI. *Minds and Machines*. 22(4): 299–324.

173 Ibid.

174 Obviously, this would require huge computational resources. Let's bracket this issue for the moment.

175 See Gray, Sidney. 2015. How a Superintelligent AI Could Convince You That You're a Simulation. *Motherboard*. URL: http://motherboard.vice.com/read/the-superintelligent-ai-says-youre-just-a-daydream.

176 Yudkowsky, Eliezer. 2002. The AI-Box Experiment. URL: http://yudkowsky.net/singularity/aibox/.

177 For Yudkowsky's IQ, see michaelgrahamrichard. 2010. Eliezer Yudkowsky—Less Wrong Q&A (4/30). YouTube. URL: https://www.youtube.com/watch?v=9eWvZLYcous.

178 Yampolskiy, Roman. 2016. Taxonomy of Pathways to Dangerous Artificial Intelligence. AI, Ethics, and Society: Technical Report. URL: http://www.aaai.org/ocs/index.php/WS/AAAIW16/paper/view/12566/12356.

179 For an excellent survey of the possible responses to superintelligence, see Sotala, Kaj, and Roman Yampolskiy. 2014. Responses to Catastrophic AGI risk: A Survey. *Physica Scripta*. 90(1). See also Barrett, Anthony, and Seth Baum. 2016. A Model of Pathways to Artificial Superintelligence Catastrophe for Risk and Decision Analysis. *Journal of Experimental and Theoretical Artificial Intelligence*. 2: 397–414. Furthermore, for an extensive list of AI guidelines, see the Asilomar AI Principles that were established during the 2017 Asilomar conference. URL: https://futureoflife.org/ai-principles/.

180 Bostrom, Nick. 2014. *Superintelligence: Paths, Dangers, Strategies*. Oxford: Oxford University Press. See also Armstrong, Stuart. 2014. *Smarter Than Us: The Rise of Machine Intelligence*. Berkeley, CA: Machine Intelligence Research Institute; and Barrat, James. 2013. *Our Final Invention: Artificial Intelligence and the End of the Human Era*. New York, NY: St. Martin's Press. Furthermore, readers can find a helpful defense of Bostrom's theses in Dafoe, Allan, and Stuart Russell. 2016. Yes, We Are Worried About the Existential Risk of Artificial Intelligence. *MIT Technology Review*. URL: https://www.technologyreview.com/s/602776/yes-we-are-worried-about-the-existential-risk-of-artificial-intelligence/.

181 Thanks to Alexey Turchin for this idea.

182 BBC. 2010. Stephen Hawking Warns over Making Contact with Aliens. BBC. URL: http://news.bbc.co.uk/2/hi/8642558.stm.

183 Quoted in Brin, David. 2013. Calling All Flash Mobs! Defend the Planet from Noisy Fools! *Contrary Brin*. URL: http://davidbrin.blogspot.com/2013/06/calling-all-flash-mobs-defend-planet.html.

184 Ibid.

185 To be clear, this quote comes from a discussion of Ludwig Boltzmann's work on probability and the second law of thermodynamics. See Price, Huw. 1996. *Time's Arrow and Archimedes' Point: New Directions for the Physics of Time*. Oxford: Oxford University Press.

186 Merton, Robert. 1936. The Unanticipated Consequences of Purposive Social Action. *American Sociological Review*. 1(6): 894–904.

187 For example, Persson and Savulescu write that "all of these problems and others like them are 'collateral damage' of a technological advance, which has promoted a boost of the living standard and an explosion of the human population. In contrast to the problems considered [else-

where], *they do not arise because of the malice or derangement of a smaller number of agents, but because of the selfish and short-sighted behaviour of masses of people*. We shall be particularly concerned with these problems to the (considerable) extent that they are caused by the behaviour of the majority of citizens of affluent liberal democracies (though in the future they might to a greater extent be caused by developing countries because of their population growth and economical growth)." Persson, Ingmar, and Julian Savulescu. 2012. *Unfit for the Future: The Need for Moral Enhancement*. Oxford: Oxford University Press.

188 See section 2.3.

189 See Schlanger, Zoë. 2014. Bioterror Lab Accidents Happen Far More Often Than We Thought. *Newsweek*. URL: http://www.newsweek.com/bioterror-lab-accidents-happen-far-more-often-we-thought-265334.

190 Cirincione, Joseph. 2013. *Nuclear Nightmares: Securing the World Before It Is Too Late*. New York, NY: Columbia University Press.

191 See Starr, Barbara. 2007. Air Force Investigates Mistaken Transport of Nuclear Warheads. CNN. URL: http://www.cnn.com/2007/US/09/05/loose.nukes/index.html?_s=PM:US.

192 See Schlosser, Eric. 2013. *Command and Control: Nuclear Weapons, the Damascus Accident, and the Illusion of Safety*. New York, NY: Penguin Books.

193 See Torres, Phil. 2016. *The End: What Science and Religion Tell Us about the Apocalypse*. Charlottesville, VA: Pitchstone Publishing.

194 Eliezer Yudkowsky appears to agree with this point. To complete the excerpted quote opening section 4.3, he writes, "All else being equal, not many people would prefer to destroy the world. Even faceless corporations, meddling governments, reckless scientists, and other agents of doom require a world in which to achieve their goals of profit, order, tenure, or other villainies. If our extinction proceeds slowly enough to allow a moment of horrified realization, the doers of the deed will likely be quite taken aback on realizing that they have actually destroyed the world. Therefore I suggest that if the Earth is destroyed, it will probably be by mistake." See Yudkowsky, Eliezer. 2008. Cognitive Biases Potentially Affecting Judgement of Global Risks. In Nick Bostrom and Milan Ćirković (editors), *Global Catastrophic Risks*. Oxford: Oxford University Press.

195 Consider this: What would it take to intentionally destroy the world? Both the *means* and the *motivation*. What would it take to accidentally destroy the world? Just the *means*. Thus, there is, *in a sense*, less that has to be the case for the agential error.

196 The type-token distinction is this: tokens are instances of a type. Consider the word "existential." This contains eight different types of letters, namely, "e," "x," "i," "s," "t," "n," "a," and "l." With respect to the letter "e," the word contains two tokens. The same goes for "i" and "t." Or, more concretely, you along with 7.5 billion others are a token of the type "*Homo sapiens*"—which itself is, in a larger taxonomic hierarchy, a token of the type "mammal," and so on.

197 Personal communication.

198 Pew Research Center. The Future of World Religions: Population Growth Projections, 2010–2050. Accessed on April 6, 2017. URL: http://www.pewforum.org/2015/04/02/religious-projections-2010-2050/.

199 Rees, Martin. 2003. *Our Final Hour: A Scientist's Warning. How Terror, Error, and Environmental Disaster Threaten Humankind's Future in This Century—on Earth and Beyond*. New York, NY: Basic Books.

200 Quoted in Wittes, Benjamin, and Gabriella Blum. 2015. *The Future of Violence: Robots and Germs, Hackers and Drones—Confronting a New Age of Threat*. New York, NY: Basic Books.

201 More specifically, as of 2008 the average person had a 1 in 250,000,000, or 0.000000004, chance of dying from a coconut falling out of a tree. See also Torres, Phil. 2016. Who Would Destroy the World? *Bulletin of the Atomic Scientists*. URL: http://thebulletin.org/who-would-destroy-world10253.

202 Müller, Vincent, and Nick Bostrom. 2014. Future Progress in Artificial Intelligence: A Survey of Expert Opinion. In Vincent Müller (editor), *Fundamental Issues of Artificial Intelligence*. Berlin: Springer. This is not the only survey on the topic, though, e.g., see also Baum, Seth, Ben Goertzel, and Ted Goertzel. 2011. How Long Until Human-Level AI? Results from an Expert Assessment. *Technological Forecasting and Social Changes*. 78(1): 185–195.

203 For a helpful discussion of AI predictions, plus a model for testing the assumptions made by such predictions (called the "counterfactual resiliency check"), see Armstrong, Stuart, Kaj Sotala, and Seán ÓhÉigeartaigh. 2014. The Errors, Insights and Lessons of Famous AI Predictions—and What They Mean for the Future. *Journal of Experimental and Theoretical Artificial Intelligence*. 26(3): 317–342.

204 Flannery, Frances. 2016. *Understanding Apocalyptic Terrorism: Countering the Radical Mindset*. New York, NY: Routledge.

205 See Ackerman, Gary. 2003. Beyond Arson? A Threat Assessment of the Earth Liberation Front. *Terrorism and Political Violence*. 15(4): 143–170.

206 See Juergensmeyer, Mark. 2017. Radical Religious Responses to Global Catastrophe. In Richard Falk, Manoranjan Mohaty, and Victor Faessel (editors), *Exploring Emerging Global Thresholds: Toward 2030*. Hyderabad, India: Orient BlackSawn, and Juergensmeyer, Mark. 2003. *Terror in the Mind of God*. Berkeley: University of California Press. Parts of this section are excerpted from Torres, Phil. 2016. Apocalypse. . . When? It Matters Which Trend Lines One Follows: Why Terrorism *Is* an Existential Threat. *Free Inquiry*. URL: https://goo.gl/WKYo4h, and Torres, Phil. 2016. Agential Risks: A Comprehensive Introduction. *Journal of Evolution and Technology*. 26(2): 31–47.

207 Juergensmeyer, Mark. 2017. Radical Religious Responses to Global Catastrophe. In Richard Falk, Manoranjan Mohaty, and Victor Faessel (editors), *Exploring Emerging Global Thresholds: Toward 2030*. Hyderabad, India: Orient BlackSawn.

208 Brennan, John. 2015. Brennan Delivers Remarks at the Center for Strategic & International Studies Global Security Forum 2015. URL: https://www.cia.gov/news-information/speeches-testimony/2015-speeches-testimony/brennan-remarks-at-csis-global-security-forum-2015.html. Italics added.

209 Hagel, Chuck. 2014. The Department of Defense Must Plan for the National Security Implications of Climate Change. URL: https://www.whitehouse.gov/blog/2014/10/13/defense-department-must-plan-national-security-implications-climate-change.

210 Department of Defense. 2015. DoD Releases Report on Security Implications of Climate Change. URL: http://www.defense.gov/News/Article/Article/612710.

211 Kelley, Colin et al. 2015. Climate Change in the Fertile Crescent and Implications of the Recent Syrian Drought. *Proceedings of the National Academy of Sciences*. 112(11): 3241–3246.

212 For an overview of these claims, see Torres, Phil. 2015. Is This How World War III Begins? Religion, End Times, Terror and the Frightening New Middle East Tinderbox. *Salon*. URL: http://www.salon.com/2015/12/20/is_this_how_world_war_iii_begins_religion_end_times_terror_and_the_frightening_new_middle_east_tinderbox/.

213 Quoted in Holthaus, Eric. 2015. New Study Says Climate Change Helped Spark Syrian Civil War. *Slate*. URL: http://www.slate.com/blogs/future_tense/2015/03/02/study_climate_change_helped_spark_syrian_civil_war.html.

214 That is, out of 9.3 billion people in total. See Pew Research Cen-

ter. The Future of World Religions: Population Growth Projections, 2010–2050. Accessed February 5, 2017. URL: http://www.pewforum. org/2015/04/02/religious-projections-2010-2050/. Although the demographic of religiously unaffiliated individuals is quickly growing in the Western world, it is shrinking as a percentage of the global population. As the director of religion research at the Pew Research Center, Alan Cooperman, puts it, "You might think of this in shorthand as the secularizing West versus the rapidly growing rest." See Pew Research Center. 2015. Event: The Future of World Religions. URL: http://www.pewforum.org/2015/04/23/live-event-the-future-of-world-religions/.

215 The Plague of Cyprian also filled many Christians at the time with "unimaginable joy" because they interpreted the plague as indicating that the world's end was near. As Cyprian himself wrote, "The kingdom of God, most beloved brothers, has begun to be imminent. The reward of life and the joy of eternal salvation, and everlasting gladness, and the gaining possession of a paradise once lost are now coming with the passing of the world." Quoted in Brent, Allen. 2010. *Cyprian and Roman Carthage*. Cambridge: Cambridge University Press.

216 See Juergensmeyer, Mark. 2017. Radical Religious Responses to Global Catastrophe. In Richard Falk, Manoranjan Mohaty, and Victor Faessel (editors), *Exploring Emerging Global Thresholds: Toward 2030*. Hyderabad, India: Orient BlackSawn.

217 This is why I have previously argued that "given the apocalyptic turn of the 2000s and the societal instability caused by environmental degradation, the Islamic State could be a mere *preview* of the sort of apocalyptically-driven terrorism that we should expect in the coming decades." See Torres, Phil. 2017. The Apocalyptic Turn and the Future of Terrorism. *Medium*. URL: https://medium.com/@philosophytorres/the-apocalyptic-turn-and-the-future-of-terrorism-f58a3ffaf63d.

218 Funk, Cary, Brian Kennedy, and Elizabeth Podrebarac Sciupac. US Public Wary of Biomedical Technologies to "Enhance" Human Abilities. Accessed February 5, 2017. URL: http://www.pewinternet.org/2016/07/26/u-s-public-wary-of-biomedical-technologies-to-enhance-human-abilities/.

219 Indeed, the challenge to human nature is one reason that the political scientist Francis Fukuyama once risibly called transhumanism the world's most dangerous idea. See Bostrom, Nick. 2005. Transhumanism: The World's Most Dangerous Idea? *Foreign Policy*. URL: http://www.nickbostrom.com/papers/dangerous.html. See also section 1.7.

220 See Cook, David. 2011. Messianism in the Shiite Crescent. Hudson

Institute. URL: http://www.hudson.org/research/7906-messianism-in-the-shiite-crescent.

221 Ibid.

222 Flannery, Frances. 2016. *Understanding Apocalyptic Terrorism: Countering the Radical Mindset*. New York, NY: Routledge.

223 Anti-Defamation League. Extremists Look to April Anniversaries. Accessed February 5, 2017. URL: http://www.adl.org/combating-hate/domestic-extremism-terrorism/c/extremists-look-to-april.html. (A paragraph break was deleted.)

Chapter 5

1 Bostrom, Nick. 2003. Are You Living in a Computer Simulation? *Philosophical Quarterly*. 53(211): 243–255.

2 Following the *We Love Philosophy* blog, we could also label (1) the "Doom Hypothesis" and (2) the "Boredom Hypothesis," since the first entails human extinction whereas the latter could occur if "most (an overwhelmingly high statistical majority) civilizations that get to this advanced computational stage [of technological maturity] see no compelling reason to run such ancestor simulations." See Julien, Alec. 2012. Are We Living in a Computer Simulation? We Love Philosophy. URL: https://welovephilosophy.com/2012/10/04/are-we-living-in-a-computer-simulation/.

3 Bostrom, Nick. 2003. Are You Living in a Computer Simulation? *Philosophical Quarterly*. 53(211): 243–255.

4 George Church notes the following: "It seems very likely that we would refrain from having sims—or least having more sims than non-sim[s], no matter how much computing power that we have. For example, we already have enormous computing power but tend to use it for discovery science, engineering, survival, novelty and pleasure seeking—with only a truly tiny fraction for any kind of simulation." Personal communication. I have, in fact, argued elsewhere that it would very likely be *deeply immoral* to run simulations that include sentient beings capable of suffering, for reasons related to the "argument from evil" and Bostrom's notion of a "mind crime." Nonetheless, I have spoken to several computer scientists who have affirmed that *if* the opportunity arose to run high-resolution ancestral simulations, they would most definitely do precisely this. So I am skeptical that humanity would refrain from running a large number of simulations in the future.

5   Bostrom, Nick. 2003. Are You Living in a Computer Simulation? *Philosophical Quarterly*. 53(211): 243–255.

6   Bostrom, Nick. The Simulation Argument FAQ. Version 1.10. Accessed on April 6, 2017. URL: http://www.simulation-argument.com/faq.html.

7   For an interesting exploration of related ideas, see Hanson, Robin. 2001. How to Live in a Simulation. *Journal of Evolution and Technology*. 7. URL: http://www.jetpress.org/volume7/simulation.htm?version=meter +at+null&module=meter-Links&pgtype=article&contentId=&mediaI d=&referrer=&priority=true&action=click&contentCollection=meter-links-click.

8   As Max Tegmark says, "My advice is to go out and do really interesting things so the simulators don't shut you down." Quoted in Moskowitz, Clara. 2016. Are We Living in a Computer Simulation? *Scientific American*. URL: https://www.scientificamerican.com/article/are-we-living-in-a-computer-simulation/.

9   Thanks to Alexey Turchin for this interesting idea.

10 See Torres, Phil. 2016. *The End: What Science and Religion Tell Us about the Apocalypse*. Charlottesville, VA: Pitchstone Publishing; Torres, Phil. 2016. Our Simulated Universe Is Just One Piece of a Matryoshka Doll of Annihilation. *Motherboard*. URL: http://motherboard.vice.com/read/ why-you-dont-want-to-live-in-a-simulation-a-response-to-elon-musk; and Torres, Phil. 2014. Why Running Simulations May Mean the End Is Near. *Ethical Technology*. URL: http://ieet.org/index.php/IEET/more/ torres20141103.

11 Indeed, Bostrom presents this fact as an argument against the "multilevel hypothesis." See Bostrom, Nick. 2003. Are You Living in a Computer Simulation? *Philosophical Quarterly*. 53(211): 243–255.

12 For my own critique of the argument, see Torres, Phil. 2017. A Potential Complication, or Problem, for the Simulation Argument. X-Risks Institute Technical Report. URL: https://goo.gl/43FWu1.

13 Indeed, most of contemporary science consists of patently "fantastical" claims about the nature and workings of reality. Consider the proposition that one species can evolve into another or that space and time form a continuum. What matters isn't how crazy an idea sounds to epistemically naive ears, but the extent to which that idea is positively supported by the totality of available and intersubjectively verifiable evidence.

14 Bulletin of the Atomic Scientists. 2015. It Is 3 Minutes to Midnight. URL: http://thebulletin.org/clock/2015. Italics added.

15 Italics added. See Torres, Phil. 2016. Three Minutes before Midnight: An Interview with Lawrence Krauss about the Future of Humanity. *Free Inquiry*. 36(4). URL: https://www.secularhumanism.org/index.php/articles/7978. Italics added.

16 For more on the policy issues surrounding superintelligence, see Bostrom, Nick, Allan Dafoe, and Carrick Flynn. 2016. Policy Desiderata in the Development of Machine Superintelligence. Working paper. URL: http://www.nickbostrom.com/papers/aipolicy.pdf.

17 As the AI risk expert Roman Yampolskiy tells the Future of Life Institute, "Weaponized AI is a weapon of mass destruction and an AI arms race is likely to lead to an existential catastrophe for humanity." Conn, Ariel. 2017. Is an AI Arms Race Inevitable? Future of Life Institute. URL: https://futureoflife.org/2017/03/09/ai-arms-race-principle/.

18 A citation was removed from this quote. See Phoenix, Chris, and Mike Treder. 2008. Nanotechnology as global catastrophic risk. In Nick Bostrom and Milan Ćirković (editors), *Global Catastrophic Risks*. Oxford: Oxford University Press.

19 Bostrom, Nick. 2005. What Is a Singleton? *Linguistic and Philosophical Investigations*. 5(2): 48–54.

20 Note again the specific meaning of "Islamism," which is not synonymous with "Islam." See endnote 77 in chapter 4 for disambiguation.

21 In fact, there is surprisingly high support for such punishments among contemporary Muslim populations. For example, "Majorities of Muslims in Egypt, Jordan, Pakistan and Nigeria say they would favor making harsh punishments such as stoning people who commit adultery; whippings and cutting off of hands for crimes like theft and robbery; and the death penalty for those who leave the Muslim religion the law in their country." Pew Research Center. 2010. Muslim Publics Divided on Hamas and Hezbollah. URL: http://www.pewglobal.org/2010/12/02/muslims-around-the-world-divided-on-hamas-and-hezbollah/.

22 For more on these values, see Bostrom, Nick. 2005. Transhumanist Values. *Review of Contemporary Philosophy*. 4: 3–14. Note also that, although I use it, "progress" is a highly problematic term that deserves an entire chapter dedicated to clarifying its many senses. (This is why in section 2.3 I put the words "civilized" and "primitive" in scare quotes.)

23 All quotes from Pinker, Steven. 2011. *The Better Angels of Our Nature: Why Violence Has Declined*. New York, NY: Penguin Books.

24 Note that a global 2013 poll found that the world overwhelmingly identifies the United States—not Iran, North Korea, or Russia—as the *num-*

ber one threat to world peace. For details, see Win/Gallup International. 2013. WIN/Gallup International. 2013. WIN/Gallup International's Annual Global End of Year Survey Shows a Brighter Outlook for 2014. URL: http://www.wingia.com/web/files/services/33/file/33.pdf?1490145172.

25   Ibid.

26   These quotes are from ibid.

27   Ibid. Yet another case worth mentioning involves Richard Nixon, who was known to drink to the point of inebriation—at times it was impossible to wake him at night. As the BBC journalist Anthony Summers writes, "At Key Biscayne, according to a Secret Service source, Nixon once lost his temper during a conversation about Cambodia. 'He just got pissed,' the agent quoted eyewitnesses as saying. 'They were half in the tank, sitting around the pool drinking. And Nixon got on the phone and said: 'Bomb the shit out of them!' 'If the president had his way,' [Henry] Kissinger growled to aides more than once, 'there would be a nuclear war each week!' This may not have been an idle jest. The CIA's top Vietnam specialist, George Carver, reportedly said that in 1969, when the North Koreans shot down a US spy plane, 'Nixon became incensed and ordered a tactical nuclear strike. . . . The Joint Chiefs were alerted and asked to recommend targets, but Kissinger got on the phone to them. They agreed not to do anything until Nixon sobered up in the morning.' The allegation of flirting with nuclear weaponry is not an isolated one. Nixon had been open to the use of tactical nuclear weapons in Vietnam as early as 1954 and as president-elect in 1968 had talked of striking 'a blow that would both end the war and win it.' A Kissinger aide who moved over to the White House, David Young, told a colleague 'of the time he was on the phone [listening] when Nixon and Kissinger were talking. Nixon was drunk, and he said, 'Henry, we've got to nuke them.''" Summers, Anthony. 2000 *The Arrogance Of Power: The Secret World Of Richard Nixon*. New York, NY: Viking Penguin.

28   Alda, Alan. 1984. What Every Woman Should Know about Men. In Martha Rainbolt and Janet Fleetwood (editors), *On the Contrary: Essays by Men and Women*. Albany, NY: State University of New York Press.

29   See Pearce, David. 2012. Transhumanism and the Abolitionist Project. URL: https://www.hedweb.com/transhumanism/2012-interview.html. Incidentally, there is some evidence that the general intelligence, or IQ, of groups is nontrivially enhanced by women. For example, researchers have found that the number of female members of a group constitutes the *one and only* factor that strongly correlates with the overall IQ of that group. If true, this has direct implications for existential risk stud-

ies, since the mitigation of existential risks must involve, on many levels, a group effort. For a more detailed discussion of this idea, see Torres, Phil. 2016. The Collective Intelligence of Women Could Save the World. Future of Life Institute. URL: https://futureoflife.org/2016/06/13/collective-intelligence-of-women-save-world/.

30 For more on bad governance and its causes, I recommend Bueno de Mesquita, Bruce, and Alastair Smith. 2011. *The Dictator's Handbook: Why Bad Behavior Is Almost Always Good Politics*. New York, NY: PublicAffairs.

31 Leslie, John. 1996. *The End of the World: The Science and Ethics of Human Extinction*. New York, NY: Routledge.

32 Sandberg, Anders, Jason Matheny, and Milan Ćirković. 2008. How Can we Reduce the Risk of Human Extinction? *Bulletin of the Atomic Scientists*. URL: http://thebulletin.org/how-can-we-reduce-risk-human-extinction.

33 Bostrom, Nick. 2013. Existential Risk Prevention as Global Priority. *Global Policy*. 4(1): 15–31.

34 See Clarke's third law, which states that "any sufficiently advanced technology is indistinguishable from magic." Clarke's Three Laws. Wikipedia. Accessed on March 27, 2017. URL: https://en.wikipedia.org/wiki/Clarke%27s_three_laws.

35 See Law, Stephen. 2010. The Evil-God Challenge. *Religious Studies*. 46(3): 353–373.

36 I also discuss this possibility in Torres, Phil. 2016. *The End: What Science and Religion Tell Us about the Apocalypse*. Charlottesville, VA: Pitchstone Publishing.

Chapter 6

1 See also endnote 13.

2 Rees, Martin. 2003. *Our Final Hour: A Scientist's Warning. How Terror, Error, and Environmental Disaster Threaten Humankind's Future in This Century—on Earth and Beyond*. New York, NY: Basic Books.

3 Bostrom, Nick. 2009. The Future of Humanity. In Jan-Kyrre Berg Olsen, Evan Selinger, and Søren Riis (editors), *New Waves in Philosophy of Technology*. New York, NY: Palgrave Macmillan.

4 Bostrom, Nick. 2013. Existential Risk Prevention as Global Priority. *Global Policy*. 4(1): 15–31.

5   Bostrom, Nick. 2008. Letter from Utopia. *Studies in Ethics, Law, and Technology.* 2(1): 1–7.

6   See, e.g., endnote 521.

7   See Torres, Phil. 2015. Are We Passing Through a Bottleneck, or Will the Explosion of Existential Risks Continue? *Ethical Technology.* URL: http://ieet.org/index.php/IEET/more/torres20150125; and Torres, Phil. 2016. *The End: What Science and Religion Tell Us about the Apocalypse.* Charlottesville, VA: Pitchstone Publishing.

8   Alternatively, focusing on agential error rather than terror, one could characterize it as a mismatch between *human fallibility*, on the one hand, and the *power of advanced technologies*, on the other.

9   Tegmark, Max. 2016. The Wisdom Race Is Heating Up. Edge.org. URL: https://www.edge.org/response-detail/26687.

10 Michio Kaku touches on both the bottleneck hypothesis and the "two trends" idea discussed by Tegmark as well as Pinker (in endnote 10). Kaku argues that *within the next century*, human civilization could transition to a Type I civilization on the Kardashev scale (discussed in endnote 85). To quote him at length: "Now, by the time you reach Type II, you are immortal. Nothing known to science can destroy a Type II civilization. Comets, meteors, earthquakes, even a supernova—a Type II civilization would be able to survive even a supernova. The danger is the transition between Type 0 and Type I, and that's where we are today. We are a Type 0 civilization. We get our energy from dead plants: oil and coal. But if you get a calculator, you can calculate when we will attain Type I status. The answer is, in about 100 years we will become planetary. We'll be able to harness all the energy output of the planet Earth. We'll play with the weather. Earthquakes, volcanoes, anything planetary we will play with. *The danger period is now.* Because we still have the savagery. We still have all the passions. We have all the sectarian fundamentalist ideas circulating around. But we also have nuclear weapons. We have chemical, biological weapons capable of wiping out life on Earth. So, I see two trends in the world today. The first trend is toward a multicultural, scientific, tolerant society. . . . However, . . . I also see the opposite trend as well. What is terrorism? Terrorism in some sense is a reaction against the creation of a Type I civilization. . . . What they're reacting to is the fact that we're heading toward a multicultural, tolerant, scientific society. And that's what they don't want. . . . Now which tendency will win? I don't know. But I hope that we emerge as a Type I civilization." Kaku, Michio. 2011. Will Mankind Destroy Itself? Big Think. URL: https://www.youtube.com/watch?v=7NPC47qMJVg.

11   See Bostrom, Nick. 2013. Existential Risk Prevention as Global Priority. *Global Policy*. 4(1): 15–31.

12   Hanson, Robin. 2016. *The Age of Em: Work, Love, and Life When Robots Rule the Earth*. Oxford: Oxford University Press.

13   For more on the term "interdiscipline," see the postscript.

14   This quote comes from Soares, Nate. 2017. Ensuring Smarter-Than-Human Intelligence Has a Positive Outcome. Talks at Google. URL: https://www.youtube.com/watch?v=dY3zDvoLoao.

15   Beckstead, Nick. 2013. A Proposed Adjustment to the Astronomical Waste Argument. LessWrong. URL: http://lesswrong.com/lw/hjb/a_proposed_adjustment_to_the_astronomical_waste/.

16   Quoted in ibid.

17   Beckstead, Nick. 2013. On the Overwhelming Importance of Shaping the Far Future. PhD diss., Rutgers University. URL: https://docs.google.com/viewer?a=v&pid=sites&srcid=ZGVmYXVsdGRvbWFpbnxuYmVVja3N0ZWFkfGd4OjExNDBjZTcwNjMxMzRmZGGE.

18   Beckstead, Nick. 2013. A Proposed Adjustment to the Astronomical Waste Argument. LessWrong. URL: http://lesswrong.com/lw/hjb/a_proposed_adjustment_to_the_astronomical_waste/.

19   Beckstead, Nick. 2013. On the Overwhelming Importance of Shaping the Far Future. PhD diss., Rutgers University. URL: https://docs.google.com/viewer?a=v&pid=sites&srcid=ZGVmYXVsdGRvbWFpbnxuYmVVja3N0ZWFkfGd4OjExNDBjZTcwNjMxMzRmZGGE.

20   Ibid.

21   Quoted from Torres, Phil. 2016. *The End: What Science and Religion Tell Us about the Apocalypse*. Charlottesville, VA: Pitchstone Publishing.

22   This definition is perhaps too anthropocentric. Cognitive enhancement is also attainable by machine intelligence, as discussed in subsection 4.3.1.

23   This enumeration borrows from Bostrom, Nick, and Anders Sandberg. 2009. Cognitive Enhancement: Methods, Ethics, Regulatory Challenges. *Science and Engineering Ethics*. 15: 311–341.

24   Although Susan Schneider points out that we should count any technology that provides a qualitatively new type cognitive capacity as an "enhancement" too, not just those that augment capacities we already have (personal communication). By analogy, consider the human sensorium. Whereas the optical telescope has enhanced a sensory modality *native*

to the human organism, namely, vision, the radio telescope has enabled humans to "see" novel features of the universe by detecting *radio waves* that are otherwise undetectable to us. In this sense, radio telescopes have given humanity a novel perceptual capacity, namely, the capacity to "perceive" electromagnetic radiation outside the narrow spectrum of visible light.

25 Bostrom, Nick, and Anders Sandberg. 2009. Cognitive Enhancement: Methods, Ethics, Regulatory Challenges. *Science and Engineering Ethics.* 15:311–341.

26 Ibid.

27 Sandberg, Anders. 2011. Cognition Enhancement: Upgrading the Brain. In Julian Savulescu, Ruud ter Meulen, and Guy Kahane (editors), *Enhancing Human Capacities.* Oxford: Blackwell Publishing.

28 Turner, D.C. et al. 2002. Cognitive Enhancing Effects of Modafinil in Healthy Volunteers. *Psychopharmacology.* 165(3): 260–269.

29 See See Nixey, Catherine. 2010. Are "Smart Drugs" Safe for Students? *Guardian.* URL: http://www.theguardian.com/education/2010/apr/06/students-drugs-modafinil-ritalin; and Bird, Steve. 2013. The Dangers for Students Addicted to Brain Viagra: Drugs Claimed to Boost Your Intellect Are Sweeping Universities—but at What Cost? *Daily Mail.* URL: http://www.dailymail.co.uk/ health/article-2451586/More-students-turning-cognitive-enhancing-drug-Modafinil-hope-boosting-grades-job-prospects.html.

30 On this point, see also Clark, Andy, and David Chalmers. 1998. The Extended Mind. *Analysis.* 58: 10–23.

31 See Muoio, Danielle. 2017. Elon Musk May Be Gearing Up for His Strangest Announcement Yet on Artificial Intelligence. *Business Insider.* URL: http://www.businessinsider.com/elon-musk-teases-neural-lace-announcement-2017-1. Some wording in the relevant paragraph comes from this article, just cited.

32 For more, see Torres, Phil. 2016. *The End: What Science and Religion Tell Us about the Apocalypse.* Charlottesville, VA: Pitchstone Publishing.

33 Again, see Bostrom, Nick, and Anders Sandberg. 2009. Cognitive Enhancement: Methods, Ethics, Regulatory Challenges. *Science and Engineering Ethics.* 15: 311–341.

34 Brady, Scott et al. 2006. *Basic Neurochemistry: Molecular, Cellular, and Medical Aspects.* 7th ed. London: Elsevier.

35 As Bostrom writes about transhumanism and eugenics, "Eugenics in

the narrow sense refers to the pre-WWII movement in Europe and the United States to involuntarily sterilize the "genetically unfit" and encourage breeding of the genetically advantaged. These ideas are entirely contrary to the tolerant humanistic and scientific tenets of transhumanism. In addition to condemning the coercion involved in such policies, transhumanists strongly reject the racialist and classist assumptions on which they were based, along with the notion that eugenic improvements could be accomplished in a practically meaningful timeframe through selective human breeding." For further discussion, see Bostrom, Nick. 2003. The Transhumanist FAQ—A General Introduction, Version 2.1. URL: http://www.nickbostrom.com/views/transhumanist.pdf.

36 Shulman, Carl, and Nick Bostrom. 2014. Embryo Selection for Cognitive Enhancement: Curiosity or Game-changer? *Global Policy*. 5(1): 85–92.

37 Bostrom, Nick. 2002. Existential Risks: Analyzing Human Extinction Scenarios and Related Hazards. *Journal of Evolution and Technology*. 9(1). To be clear, Bostrom is referring not just to cognitive enhancements, which he refers to as "intelligence augmentation," but to "information technology" and "surveillance" as well.

38 Note here that insofar as terrorists are educated, they are often educated in fields like religious studies and engineering. Neither of these fields is notable for *strongly* encouraging, fostering, or cultivating critical thinking, empathy, sympathetic concern, intellectual reflection, fallibilism, and so on, which is (one reason) why the humanities remains important and relevant, however "ornamental" it may appear to those who prefer "instrumental" fields.

39 For more on the connection between literature and empathy, see Pinker, Steven. 2011. *The Better Angels of Our Nature: Why Violence Has Declined*. New York, NY: Penguin Books.

40 Harris, John. 2007. *Enhancing Evolution: The Ethical Case for Making Better People*. Princeton, NJ: Princeton University Press.

41 Incidentally, Harris also offers a vigorous argument *against* moral bioenhancement, which he believes could compromise human freedom. See Harris, John. 2011. Moral Enhancement and Freedom. *Bioethics*. 25(2): 102–111.

42 See, e.g., Kanazawa, Satoshi. 2010. Why Liberals and Atheists Are More Intelligent. *Social Psychology Quarterly*. 73(1): 33–57.

43 See, e.g., Barber, Nigel. 2010. The Real Reason Atheists Have Higher IQs. *Psychology Today*. URL: https://www.psychologytoday.com/blog/

the-human-beast/201005/the-real-reason-atheists-have-higher-iqs.

44 Gervais, Will, and Ara Norenzayan. 2012. Analytic Thinking Promotes Religious Disbelief. *Science*. 336: 493.

45 Quoted in Attwood, Rebecca. 2008. High IQ Turns Academics into Atheists. *Times Higher Education*. URL: https://www.timeshighereduca-tion.com/news/high-iq-turns-academics-into-atheists/402381.articlc.

46 I am here ignoring the potential upsides of religious culture because the focus is direct strategies for reducing existential risk.

47 Pinker, Steven. 2011. *The Better Angels of Our Nature: Why Violence Has Declined*. New York, NY: Penguin Books.

48 Ibid. Pinker calls this the "moral Flynn effect."

49 See Ellis, Lee, Kevin Beaver, and John Wright. *Handbook of Crime Correlates*. New York, NY: Academic Press.

50 Langman, Peter. 2009. Shakespeare and School Shooters, Part 1. *Psychology Today*. URL: https://www.psychologytoday.com/blog/keeping-kids-safe/200910/shakespeare-and-school-shooters-part-1.

51 Ibid. As for the Nietzsche quote, it could be a mistranslation of the following line from *Thus Spoke Zarathustra*: "The earth, said he, hath a skin; and this skin hath diseases. One of these diseases, for example, is called 'man.'" Nonetheless, "The world is beautiful, but has a disease called Man" appears frequently in books, in articles, and on websites. It is therefore plausible that Eric Harris might have encountered it, as Peter Langman suggests, and this could have further reinforced his omnicidal persuasion that humanity must be destroyed.

52 See Maese, Rick, and Peter Hermann. 2012. Portrait of Adam Lanza and His Family Begins to Emerge. *Washington Post*. URL: https://www.washingtonpost.com/national/portrait-of-adam-lanza-and-his-family-begins-to-emerge/2012/12/17/376759ce-4862-11e2-820e-17eef-ac2f939_story.html?utm_term=.b01a58e5d1ac.

53 Perhaps some cognitive enhancements could be designed *specifically* to improve the crucial "abstract reasoning" component of intelligence that Pinker emphasizes, and perhaps *this* could mitigate the agential threat posed by Harris, Lanza, and Manson. Perhaps the lesson is that we must make sure that cognitive enhancements target the particular properties of cognition that could make people more *reasonable*.

54 Langman, Peter. 2010. Influences on the Ideology of Eric Harris. URL: https://schoolshooters.info/sites/default/files/harris_influences_ideology_1.2.pdf. See also Langman, Peter. 2009. *Why Kids Kill: Inside*

the Minds of School Shooters. New York, NY: Palgrave Macmillan.

55 To lay my cards on the table, I am sympathetic with noncognitivist me-taethical theories like emotivism. I do not believe that there are mind-independent moral facts that SNUs are simply failing to grasp.

56 Bigelow, John, and Robert Pargetter. 1999. Functions. In David J. Buller (editor), *Function, Selection, and Decision*. Albany: State University of New York Press.

57 Although Ingmar Persson argues that SNU does contain a cognitive mistake, and thus cognitive enhancements could make one less likely to espouse SNU. The mistake is that SNU sees pleasure as merely the *absence* of suffering rather than as something *positive*, which "is clearly a false view," because "the absence of suffering could be a state which in it-self is neutral or indifferent rather than good." Persson adds that even if pain can be more intense than pleasure (which may indeed be the case), "This could justify a *greater* moral concern about pain, but it could not justify being concerned *only* about the elimination of pain, etc. . . . [I]f you recognize that pleasure is a genuine component of well-being then, if you are altruistically concerned about the well-being of others, you will be concerned also that others will feel pleasure, that the net surplus of pleasure over pain be as great as possible. Thus, you will prefer total annihilation only if that can't be made the case." Personal communica-tion.

58 See Joy, Bill. 2000. Why the Future Doesn't Need Us. *Wired*. URL: https://www.wired.com/2000/04/joy-2/.

59 Indeed, he appears to have had an IQ of 167. See Chase, Alston. 2000. Harvard and the Making of the Unabomber. *Atlantic*. URL: https://www.theatlantic.com/magazine/archive/2000/06/harvard-and-the-making-of-the-unabomber/378239/. Along these lines, Frances Flannery writes that "in terms of demographics, the members of ELF [i.e., the Earth Lib-eration Front] are most often male, well educated, [and] technologically literate." Flannery, Frances. 2016. *Understanding Apocalyptic Terrorism: Countering the Radical Mindset*. New York, NY: Routledge.

60 Persson, Ingmar, and Julian Savulescu. 2008. The Perils of Cognitive Enhancement and the Urgent Imperative to Enhance the Moral Char-acter of Humanity. *Journal of Applied Philosophy*. 25(3): 162–177. This could also be problematic for reasons relating to the "first law of the ethics of technology," which states that "technology evolves at a geo-metric rate, while social policy develops at an arithmetical rate. In other words, changing societal attitudes takes a much greater time than it does for technology to evolve." Walker, Mark. 2009. H+: Ship of Fools: Why

Transhumanism Is the Best Bet to Prevent the Extinction of Civiliza-tion. *Metanexus*. URL: http://www.metanexus.net/essay/h-ship-fools-why-transhumanism-best-bet-prevent-extinction-civilization.

61   Quoted in Hauser, Robert, and Alberto Palloni. 2011. Adolescent IQ and Survival in the Wisconsin Longitudinal Study. *Journal of Gerontol-ogy: Psychological Sciences*. 66B(Suppl 1): i91–i101.

62   Gottfredson, Linda, and Ian Deary. 2004. Intelligence Predicts Health and Longevity, but Why? *Current Directions in Psychological Science*. 13(1): 1–4. A quick note about Linda Gottfredson, since I cite her twice. She holds a number of views about race and intelligence that appear not only scientifically untenable, but *morally abominable*—and, indeed, she has received hundred of thousands of dollars from a racist organization called the Pioneer Fund. (Sadly, Garrett Hardin of chapter 3 was also funded by this organization.) This has earned her the dubious honor of having an entire webpage dedicated to her problematic and unethical ideas on the Southern Poverty Law Center website. Nonetheless, there is something called the "genetic fallacy," which essentially states that the truth of a proposition is *independent of* the source of that proposition. (For example, "2 plus 2 equals 4" would not be false *simply because* Pot Pol uttered it.) Thus, with respect to the specific articles here cited, I find no arguments or conclusions that appear to be colored by prior ideological commitments to white supremacy. I therefore take these data to be as valid as any others that one would find in reputable (to my knowledge) peer-reviewed journals like *Current Directions in Psy-chological Science* and *Journal of Personality and Social Psychology*. The present author cares deeply about verifiable data, but also *strongly con-demns* the flagrant lack of intellectual integrity and moral thoughtful-ness of Gottfredson. For more, see Southern Poverty Law Center. Linda Gottfredson. Accessed on April 25, 2017. URL: https://www.splcenter.org/fighting-hate/extremist-files/individual/linda-gottfredson.

63   Gottfredson, Linda. 2004. Intelligence: Is It the Epidemiologists' Elusive "Fundamental Cause" of Social Class Inequalities in Health? *Journal of Personality and Social Psychology*. 86(1): 174–199.

64   Not to mention "moral truths" like "polluting is wrong because it de-stroys the environment, and environmental destruction causes harm to both human and nonhuman life. This is a point that Persson and Sa-vulescu emphasize in Persson, Ingmar, and Julian Savulescu. 2012. *Unfit for the Future: The Need for Moral Enhancement*. Oxford: Oxford Uni-versity Press.

65   Bostrom, Nick. 2008. Three Ways to Advance Science. *Nature* Podcast.

URL: http://www.nickbostrom.com/views/science.pdf.

66 It could also lead to a better understanding of whether or not we exist in a computer simulation. In fact, some scientists are already investigating ways to get clues, through empirical science, about whether the universe is simulated or not. For more, see American Museum of Natural History. 2016. 2016 Isaac Asimov Memorial Debate: Is the Universe a Simulation? YouTube. URL: https://www.youtube.com/watch?v=wgSZA3NPpBs.

67 See Pinker, Steven. 2011. *The Better Angels of Our Nature: Why Violence Has Declined*. New York, NY: Penguin Books.

68 Persson, Ingmar, and Julian Savulescu. 2012. *Unfit for the Future: The Need for Moral Enhancement*. Oxford: Oxford University Press.

69 Essentially, what Persson and Savulescu's proposal amounts to is making men more like women—or rather, making men more like men who are like women. It also entails making conservatives more like liberals, if the distinction between such individuals is seen through the prism of Jonathan Haidt's "moral foundations theory." Basically, a more "feminine" and liberal society is what we ought to aim for. See Haidt, Jonathan. 2012. *The Righteous Mind: Why Good People Are Divided by Politics and Religion*. New York, NY: Pantheon Books.

70 Persson, Ingmar, and Julian Savulescu. 2012. *Unfit for the Future: The Need for Moral Enhancement*. Oxford: Oxford University Press.

71 Ibid.

72 Ibid.

73 Savulescu, Julian, and Ingmar Persson. 2012. Moral Enhancement, Freedom and the God Machine. *Monist*. 95(3): 339–421.

74 Persson, Ingmar, and Julian Savulescu. 2012. *Unfit for the Future: The Need for Moral Enhancement*. Oxford: Oxford University Press.

75 Ibid.

76 Savulescu, Julian, and Ingmar Persson. 2012. Moral Enhancement, Freedom and the God Machine. *Monist*. 95(3): 339–421.

77 Persson, Ingmar, and Julian Savulescu. 2012. *Unfit for the Future: The Need for Moral Enhancement*. Oxford: Oxford University Press.

78 As the Provisional Irish Republican Army (PIRA) eerily declared after nearly killing Margaret Thatcher, "We only have to be lucky once. You have to be lucky all the time." This points to an incredibly serious offensive/defensive asymmetry that will haunt agential risk mitigation efforts in the future. Quoted in Iacono, Daniela. 1984. IRA Wing Says Bombing

Result of Britain's Ireland Policy. *United Press International*. URL: http://www.upi.com/Archives/1984/10/13/IRA-wing-says-bombing-result-of-Britains-Ireland-policy/6913466488000/. But see also Torres, Phil. 2016. Apocalypse. . . When? It Matters Which Trend Lines One Follows: Why Terrorism *Is* an Existential Threat. *Free Inquiry*. URL: https://goo.gl/WKYo4h.

79 A similar point is made by Sparrow, Robert. 2014. Egalitarianism and moral bioenhancement. *American Journal of Bioethics*. 14(4): 20–28.

80 In fact, this is what Persson and Savulescu advocate, along with mass surveillance systems, which are discussed in subsection 6.3.3.

81 Flannery, Frances. 2016. *Understanding Apocalyptic Terrorism: Countering the Radical Mindset*. New York, NY: Routledge. Italics added.

82 Just "consider the people who learn about the promise of a perfect world yet nonetheless oppose it," Pinker writes. "They [the unbelievers] are the only things standing in the way of a plan that could lead to infinite goodness. How evil are they? You do the math." Pinker, Steven. 2011. *The Better Angels of Our Nature: Why Violence Has Declined*. New York, NY: Penguin Books.

83 Furthermore, the conception of justice that motivates apocalyptic terrorism is that of *cosmic justice*, according to which God will distribute punishments and rewards based on whether individuals adhere to the one true religion. This kind of justice constitutes the ultimate *theodicy*, or vindication of evil in the world. It follows that catalyzing such a momentous event through violence should be appealing to apocalyptic warriors: the sooner God exacts justice on this evil world, sodden in suffering and sin, the better off believers will be.

84 That is, ignoring the logistical issue of getting such terrorists to use moral bioenhancements in the first place.

85 Agar, Nicholas. 2013. Moral Bioenhancement Is Dangerous. *Journal of Medical Ethics*. 41: 343–345. Note that Agar is a critic of moral bioenhancement. For example, he writes that moral bioenhancement "is perilous not because of the end that is sought, but instead because of the way that moral bioenhancers will almost certainly work. There are unlikely to be any pills or injections that directly produce in us morally superior judgments or motivations. Moral bioenhancers will achieve that end indirectly by strengthening some among the diverse collection of cognitive, emotional and motivational inputs into moral thinking. Moral bioenhancers will fail to morally enhance when they strengthen to too great a degree one or some among the diverse influences on mor-

al judgment. Unbalanced excesses in influences on moral thinking are likely results of attempts at moral enhancement by biomedical means. These resemble insufficiencies in cognitive, emotional and motivational inputs in making us less morally good. Both excesses and insufficiencies throw out the proper balance of psychological and emotional influences that informs sound moral judgment." Ibid.

86   Quoted in Langman, Peter. 2010. Influences on the Ideology of Eric Harris. URL: https://schoolshooters.info/sites/default/files/harris_influences_ideology_1.2.pdf, and Langman, Peter. 2009. Rampage school shooters: A typology. *Aggression and Violent Behavior*. 14: 79–86.

87   To borrow two lines, used in a different context, from Persson and Savulescu. See Persson, Ingmar, and Julian Savulescu. 2012. *Unfit for the Future: The Need for Moral Enhancement*. Oxford: Oxford University Press.

88   As I have written elsewhere, "For example, a negative utilitarian could find her or himself hesitant to follow-through on actions involving WTDs that she or he *believes* are moral. The biological instinct of self-preservation, or the worry that a WTD attack could fail (thereby resulting in far more suffering), could be sufficient to prevent one from acting. She or he might then acquire mostropics to surmount this reluctance." Torres, Phil. Forthcoming. Moral Bioenhancement and Agential Risks: Good and Bad Outcomes. *Bioethics*.

89   Indeed, as Frances Flannery writes, "Since REAR [the Radical Environmental and/or Animal Rights] activists espouse a biocentric view and *identify deeply* with animals and elements of the environment, they regularly experience persecution on their behalf. Violent REAR activists believe that they are fighting against an ongoing 'genocide' being perpetrated against innocent lives that cannot defend themselves; furthermore, *they feel a close kinship with those lives*." Flannery, Frances. 2016. *Understanding Apocalyptic Terrorism: Countering the Radical Mindset*. New York, NY: Routledge. Italics added. For a short introduction to Gaia theory, see Gaia Theory. Overview. Accessed on March 8, 2017. URL: http://www.gaiatheory.org/overview/.

90   Brennan, Andrew, and Yeuk-Sze Lo. 2016. Environmental Ethics. *The Stanford Encyclopedia of Philosophy*. URL: https://plato.stanford.edu/archives/win2016/entries/ethics-environmental.

91   Or, put differently, humanity has "screwed over" nature, so "nature" should reciprocate by "screwing over" humanity. Insofar as "nature" is unable to do this, ecoterrorists must take the reins to exact justice on *Homo sapiens*. Indeed, with respect to iterated prisoner's dilemma games, players who use the tit-for-tat strategy will initially cooperate

with their opponent, after which they will mirror the opponent's previous decision, meaning that if the previous decision was to defect, the other player will choose to defect as well.

92 Whether or not they should become compulsory should be based on a careful analysis of their various effects. For example, if they worsen ecoterrorism by a factor of 5 while mitigating apocalyptic terrorism by a factor of 3, then their net effect might be negative. On the other hand, if apocalyptic terrorism becomes a much greater threat than ecoterrorism, then there might be some rationale for implementing an involuntary regime. More research is needed on this crucial issue.

93 Persson, Ingmar, and Julian Savulescu. 2012. *Unfit for the Future: The Need for Moral Enhancement*. Oxford: Oxford University Press.

94 Ibid.

95 A particularly poignant expression of this basic idea comes from a letter written by a Holocaust survivor: "I am a survivor of a concentration camp. My eyes saw what no person should witness: gas chambers built by learned engineers. Children poisoned by educated physicians. Infants killed by trained nurses. Women and babies shot by high school and college graduates. So, I am suspicious of education. My request is: Help your children become human. Your efforts must never produce learned monsters, skilled psychopaths or educated Eichmanns. Reading, writing, and arithmetic are important only if they serve to make our children more human." Quoted in Ginott, Haim. 1993. *Teacher and Child: A Book for Parents and Teachers*. New York, NY: Scribner Paper Fiction. (Thanks to Gary Ackerman for apprising me of this quote.)

96 Persson, Ingmar, and Julian Savulescu. 2012. *Unfit for the Future: The Need for Moral Enhancement*. Oxford: Oxford University Press.

97 For example, as Robert Sparrow puts it, "If not enough other people do it, there is no point in my doing it; if enough other people do it, it's not in my interests to do it—I might as well free ride on the moral enhancement of others. Thus, the project of *voluntary* moral bioenhancement to prevent climate change presupposes the sense of social solidarity that it is supposed to bring about." Sparrow, Robert. 2014. Egalitarianism and Moral Bioenhancement. *The American Journal of Bioethics*. 14(4): 20–28. Italics added. It follows that, as Persson and Savulescu explicitly argue in one paper, moral bioenhancement programs must be universal and compulsory. Persson, Ingmar, and Julian Savulescu. 2008. The Perils of Cognitive Enhancement and the Urgent Imperative to Enhance the Moral Character of Humanity. *Journal of Applied Philosophy*. 25(3): 162-177.

98   Personal communication.

99   See Torres, Phil. 2016. Apocalypse Soon? How Emerging Technologies, Population Growth, and Global Warming Will Fuel Apocalyptic Terrorism in the Future. *Skeptic*. URL: https://goo.gl/rIwJkm.

100 Mental Health America. Position Statement 22: Involuntary Mental Health Treatment. Accessed on January 12, 2017. URL: http://www.mentalhealthamerica.net/positions/involuntary-treatment.

101 Persson, Ingmar, and Julian Savulescu. 2012. *Unfit for the Future: The Need for Moral Enhancement*. Oxford: Oxford University Press. Similarly, George Church writes in the context of synthetic biology threats that "there is no perfect solution, but a partial solution is to have more surveillance of whatever you can to monitor, where the expertise is going, and where the materials required for practice are going. And to discourage any kind of negative use and encourage positive uses in every way you can as a top societal priority, not something you just give lip service to for a microsecond in some Congressional session. Whatever it takes." Church, George. 2006. Constructive Biology. Edge.org. URL: https://www.edge.org/conversation/george_church-constructive-biology.

102 Bostrom, Nick. 2013. Existential Risk Prevention as Global Priority. *Global Policy*. 4(1): 15–31.

103 See Mann, Steve, Jason Nolan, and Barry Wellman. 2003. Sousveillance: Inventing and Using Wearable Computing Devices for Data Collection in Surveillance Environments. *Surveillance and Society*. 1(3): 331–355. A similar idea is the futurist Jamais Cascio's notion of a *participatory panopticon*. As he writes, "the world of the participatory panopticon is not as interested in privacy, or even secrecy, as it is in lies. A police officer lying about hitting a protestor, a politician lying about human rights abuses, a potential new partner lying about past indiscretions—all of these are harder in a world where everything might be on the record. The participatory panopticon is a world where accusations can easily be documented, where corporations will become more transparent to stakeholders as a matter of course, where officials may even be required to wear a recorder while on duty, simply to avoid situations where they are discovered to have been lying. It's a world where we can all be witnesses with perfect recall. Ironically, it's a world where trust is easy, because lying is hard." See Cascio, Jamais. 2013. Anticipatory Mythologies. Open the Future. URL: http://www.openthefuture.com/participatory_panopticon/. Furthermore, the futurist and security expert Philippe van Nedervelde "has come up with the concept of the 4 E's: 'Everyone has Eyes and Ears Everywhere,'" which could be realized

by what he calls "Panoptic Smart Dust Sousveillance" (PSDS). In Van Nedervelde's words, "Today, 'smart dust' refers to tiny MEMS [or 'micro-electro-mechanical systems'] devices nicknamed 'motes' measuring one cubic millimeter or smaller capable of autonomous sensing, computation and communication in wireless ad-hoc mesh networks. In the not too far future, NEMS [or 'nano-electro-mechanical systems'] will enable quite literal 'smart dust' motes so small—50 cubic microns or smaller—that they will be able to float in the air just like 'dumb dust' particles of similar size and create solar-powered mobile sensing 'smart clouds.'" Van Nedervelde envisages that, "the lower levels of the Earth's atmosphere [being] filled with smart dust motes at an average density of three motes per cubic yard of air. If engineered, deployed, maintained and operated by the global citizenry for the global citizenry, this would create a 'Panoptic Smart Dust Sousveillance' (PSDS) system—essentially a citizen's sousveillance network effectively giving Everyone Eyes and Ears Everywhere, and thereby effectively and efficiently realizing—or at least enabling in the sense of making possible—so-called 'reciprocal accountability' throughout civilized society." Finally, he adds that, "Assuming that most of the actual sousveillance would not be done by humans but by pattern-spotting machines instead, this would indeed be the end of what I have called 'absolute privacy'—still leaving most with, in my view acceptable, 'relative privacy'— but most probably also the end of SIMAD or other terrorist attacks as well as, for instance, the end of violence and other forms of abuse against children, women, the elderly and other victims of domestic violence and other abuse." Quoted in Diaz, Jesus. 2013. Can One Single Person Destroy the Entire World? *iO9*. URL: http://io9.gizmodo.com/could-a-single-individual-really-destroy-the-world-1471212186/1471327744.

104 Brin, David. 1996. The Transparent Society. *Wired*. URL: https://www.wired.com/1996/12/fftransparent/.

105 They define a "eucatastrophe" as "an event which causes there to be much more expected value after the event than before." They add that, "Armed with this concept, we can draw a new lesson. Just as we should strive to avoid existential catastrophes, we should also seek existential eucatastrophes." See Cotton-Barratt, Owen, and Toby Ord. 2015. Existential Risk and Existential Hope: Definitions. Future of Humanity Institute Technical Report. URL: http://www.fhi.ox.ac.uk/Existential-risk-and-existential-hope.pdf.

106 Specifically, Hawking states that, "The rise of powerful AI will be either the best or the worst thing ever to happen to humanity. We do not yet know which." From the Leverhulme Centre for the Future of Intel-

ligence website. Accessed on March 16, 2017. URL: http://lcfi.ac.uk/. See also Hern, Alex. 2016. Stephen Hawking: AI Will Be "Either Best or Worst Thing" for Humanity. *Guardian*. URL: https://www.theguardian.com/science/2016/oct/19/stephen-hawking-ai-best-or-worst-thing-for-humanity-cambridge. Incidentally, David Chalmers expresses a similar opinion, writing that "an intelligence explosion has enormous potential benefits: a cure for all known diseases, an end to poverty, extraordinary scientific advances, and much more. It also has enormous potential dangers: an end to the human race, an arms race of warring machines, the power to destroy the planet." See Chalmers, David. 2010. The Singularity: A Philosophical Analysis. *Journal of Consciousness Studies*. 17(9–10): 7–65.

107 Bostrom, Nick. 2014. *Superintelligence: Paths, Dangers, Strategies*. Oxford: Oxford University Press.

108 Furthermore, it's worth noting that there could be tension and conflict between different types of (human) agential risks, since the realization of one risky agent's ultimate goal could *permanently thwart* the realization of another's. For example, an anarcho-primitivist's aim to establish a new Paleolithic-like milieu is in direct conflict with the sadistic sociopathic misanthrope's goal of exterminating all humans on the planet. Similarly, imagine a strong negative utilitarian (SNU) with omnicidal dreams being preempted by a catastrophically violent neo-Luddite who gains access to a civilization-destroying weapon. Apocalyptic terrorists could also complicate the plans of SNUs, anti-civilization extremists, and humanity-hating sociopaths by unilaterally bringing about a catastrophe motivated by active apocalyptic beliefs concerning the will of God/Allah. Perhaps this fact could be leveraged to get different agential risks to neutralize each other—yet another neglected topic that, as such, deserves more research.

109 IAEA. Treaty on the Non-Proliferation of Nuclear Weapons (NTP). Accessed on December 14, 2016. URL: https://www.iaea.org/publications/documents/treaties/npt.

110 In fact, Sam Harris argues that "we need something like a Manhattan Project on the topic of artificial intelligence. Not to build it, because I think we'll inevitably do that, but to understand how to avoid an arms race and to build it in a way that is aligned with our interests. When you're talking about superintelligent AI that can make changes to itself, it seems that we only have one chance to get the initial conditions right, and even then we will need to absorb the economic and political consequences of getting them right." See Harris, Sam. 2016. Can We Build

AI without Losing Control over It? TED. URL: https://www.ted.com/talks/sam_harris_can_we_build_ai_without_losing_control_over_it?language=en. Similarly, the author James Barrat states that humanity needs to, "Create a global public-private partnership to ride herd on those with AGI ambitions, something like the International Atomic Energy Agency (IAEA). Until that organization is created, form a consortium with deep pockets to recruit the world's top AGI researchers. Convince them of the dangers of unrestricted AGI development, and help them proceed with utmost caution. Or compensate them for abandoning AGI dreams." Quoted in Diaz, Jesus. 2013. Can One Single Person Destroy the Entire World? *iO9*. URL: http://io9.gizmodo.com/could-a-single-individual-really-destroy-the-world-1471212186/1471327744.

111 See Joy, Bill. 2000. Why the Future Doesn't Need Us. *Wired*. URL: https://www.wired.com/2000/04/joy-2/.

112 Quoted in Stewart Brand, Kevin Kelly, and George Dyson. 2011. A *Edge* Conversation in Munich. Edge.org. URL: https://www.edge.org/documents/archive/edge338.html.

113 Walker, Mark. 2009. Ship of Fools: Why Transhumanism Is the Best Bet to Prevent the Extinction of Civilization. *Metanexus*. URL: http://www.metanexus.net/essay/h-ship-fools-why-transhumanism-best-bet-prevent-extinction-civilization.

114 See Hughes, James. 2001. Relinquishment or Regulation: Dealing with Apocalyptic Technological Threats. URL: http://www.changesurfer.com/Acad/RelReg.pdf.

115 Walker, Mark. 2009. Ship of Fools: Why Transhumanism is the Best Bet to Prevent the Extinction of Civilization. *Metanexus*. URL: http://www.metanexus.net/essay/h-ship-fools-why-transhumanism-best-bet-prevent-extinction-civilization.

116 See Winner, Langdon. 1977. *Autonomous Technology: Technics-out-of-Control as a Theme in Political Thought*. Cambridge, MA: MIT Press. See also Bostrom's "Technological Completion Conjecture" propounded in Bostrom, Nick. 2009. The Future of Humanity. In Jan-Kyrre Berg Olsen, Evan Selinger, and Soren Riis (editors), *New Waves in Philosophy of Technology*. New York, NY: Palgrave Macmillan.

117 See Kurzweil, Ray. 2005. *The Singularity Is Near: When Humans Transcend Biology*. New York, NY: Viking.

118 Ibid. Note that Freitas and Merkle say something similar when they write, "How can we avoid 'throwing out the baby with the bathwater'? The correct solution . . . starts with a *carefully targeted* moratorium or

outright legal ban on the most dangerous kinds of nanomanufacturing systems, while still allowing the safe kinds of nanomanufacturing systems to be built—subject to appropriate monitoring and regulation commensurate with the lesser risk that they pose." See Freitas, Robert, and Ralph Merkle. 2004. *Kinematic Self-Replicating Machines.* Georgetown, TX: Landes Bioscience. Italics added.

119 Bostrom, Nick. 2014. *Superintelligence: Paths, Dangers, Strategies.* Oxford: Oxford University Press.

120 Or it could make a "hard takeoff" scenario more probable. See ibid.

121 Kurzweil, Ray. 2005. *The Singularity Is Near: When Humans Transcend Biology.* New York, NY: Viking.

122 Ibid. Another problem is that the development of defensive technologies tends to lag behind the development of offensive weapons, thereby resulting in a period of heightened danger. As Bostrom writes, "While a nanotechnic defense system (which would act as a global immune system capable of identifying and neutralizing rogue replicators) appears to be possible in principle, it could turn out to be more difficult to construct than a simple destructive replicator. This could create a window of global vulnerability between the potential creation of dangerous replicators and the development 24 of an effective immune system. It is critical that nano-assemblers do not fall into the wrong hands during this period." See Bostrom, Nick. 2003. The Transhumanist FAQ—A General Introduction, Version 2.1. URL: http://www.nickbostrom.com/views/transhumanist.pdf.

123 Freitas, Robert, and Ralph Merkle. 2004. *Kinematic Self-Replicating Machines.* Georgetown, TX: Landes Bioscience.

124 Ibid.

125 Matheny, Jason. 2007. Reducing the Risk of Human Extinction. *Risk Analysis.* 27(5): 1335–1344. Some citations original to this quote have been removed. See also Tonn, Bruce. 1999. Transcending Oblivion. *Futures.* 31: 351–359.

126 Quoted in, respectively, Highfield, Roger. 2001. Colonies in Space May Be Only Hope, Says Hawking. *Telegraph.* URL: http://www.telegraph.co.uk/news/uknews/1359562/Colonies-in-space-may-be-only-hope-says-Hawking.html; and Gerken, James. 2015. Stephen Hawking Predicts Humans Won't Last Another 1,000 Years on Earth. *Huffington Post.* URL: http://www.huffingtonpost.com/2015/04/28/stephen-hawking-humanity-1000-years_n_7160870.html. Elsewhere, Hawking has stated, "I believe that the long-term future of the human race must be

space, and that it represents an important life insurance for our future survival, as it could prevent the disappearance of humanity by coloniz-ing other planets." See Lewis, Tanya. Stephen Hawking Thinks These 3 Things Could Destroy Humanity. *Live Science*. URL: http://www.li-vescience.com/49952-stephen-hawking-warnings-to-humanity.html.

127 Griffin, Michael. 2006. Quotes and Speeches in Favor of Human Space Exploration and Settlement. National Space Society. URL: http://www.nss.org/legislative/quotes/Speech_Michael_Griffin_2006_09_12.html.

128 Quoted in Anderson, Ross. 2014. Exodus. *Aeon*. URL: https://aeon.co/essays/elon-musk-puts-his-case-for-a-multi-planet-civilisation.

129 Parfit, Derek. 2017. *On What Matters*. Vol. 3 Oxford: Oxford University Press.

130 Quoted in Hersher, Rebecca, and Camila Domonoske. 2016. Elon Musk Unveils His Plan for Colonizing Mars. NPR. URL: http://www.npr.org/sections/thetwo-way/2016/09/27/495622695/this-afternoon-elon-musk-unveils-his-plan-for-colonizing-mars. Note also that there are companies currently working on designing *space elevators* that could fa-cilitate the transfer of people into space. As I have written elsewhere: "It wasn't that long ago that engineers deemed such technology impossible, primarily because a space elevator would require a *huge* cable to connect the base station, most likely located in the ocean (so it can move around to avoid inclement weather), to a counterweight at the other end, orbit-ing 'geosynchronously' above the base. No known material was strong enough to span even a fraction of this distance—that is, until the inven-tion of *carbon nanotubes*. These are, 'on an ounce-for-ounce basis . . . at least 117 times stronger than steel and 30 times stronger than Kevlar.' If long carbon nanotube strands were woven together to create a 'ribbon,' its tensile strength could be sufficient for the cable to vertically traverse Earth's atmosphere, thus making possible an elevator into space." Torres, Phil. 2016. *The End: What Science and Religion Tell Us about the Apoca-lypse*. Charlottesville, VA: Pitchstone Publishing.

131 Deudney, Daniel. Forthcoming. *Dark Skies: Space Expansionism, Plan-etary Geopolitics, and the Ends of Humanity*. Oxford: Oxford University Press.

132 Dyson, Freeman. 1979. *Disturbing the Universe*. New York, NY: Basic Books.

133 As Jakub Drmola and Miroslav Mareš put it, "Sooner or later, in order to avoid the fate of the dinosaurs, humanity needs to develop scientific and technological capabilities to prevent extinction-level impact events. But

most solutions bring about new challenges, because new technologies rarely have only one application. Here lies the dilemma: any technology allowing us to deflect asteroids from a collision trajectory with the Earth could also be used to direct them towards the Earth. This means we could potentially turn any future near-miss into an impact, with all its devastating consequences." See Drmola, Jakub, and Miroslav Mareš. 2015. Revisiting the Deflection Dilemma. *News and Reviews in Astronomy & Geophysics*. 56(5): 5.15–5.18. See also Ostro, Steven, and Carl Sagan. 1998. Cosmic Collisions and Galactic Civilizations. *Astronomy and Geophysics: The Journal of the Royal Astronomical Society*. 39(4): 22–24. Thanks to the astronomer Alan Hale for bringing this issue to my attention (personal communication).

134 This is also quoted in Torres, Phil. 2016. *The End: What Science and Religion Tell Us about the Apocalypse*. Charlottesville, VA: Pitchstone Publishing. See also Tyson, Neil deGrasse. 2012. We Can Survive Killer Asteroids—But it Won't be Easy. *Wired*. URL: https://www.wired.com/2012/04/opinion-tyson-killer-asteroids/. A single comma was removed for grammatical reasons.

135 Global Challenges Foundation. 2016. Global Catastrophic Risks 2016. URL: http://globalprioritiesproject.org/wp-content/uploads/2016/04/Global-Catastrophic-Risk-Annual-Report-2016-FINAL.pdf.

136 Ibid.

137 Ibid.

138 Falk, Dan. 2017. Can Hacking the Planet Stop Runaway Climate Change? NBC. URL: http://www.nbcnews.com/mach/environment/can-hacking-planet-stop-runaway-climate-change-n752221.

139 Hanson, Robin. 2008. Catastrophe, Social Collapse, and Human Extinction. In Nick Bostrom and Milan Ćirković (editors), *Global Catastrophic Risks*. Oxford: Oxford University Press.

140 Ibid.

141 Baum, Seth, David Denkenberger, and Jacob Haqq-Misra. 2015. Isolated Refuges for Surviving Global Catastrophes. *Futures*. 74: 45–56.

142 Note that this is directly relevant to the "last few people problem" mentioned in Box 2. Simply surviving a global catastrophe isn't enough; the remaining population must also have sufficient genetic diversity to bounce back.

143 Gorvett, Zaria. 2016. Could Just Two People Repopulate Earth? *BBC*. URL: http://www.bbc.com/future/story/20160113-could-just-two-peo-

ple-repopulate-earth; and Jebari, Karim. 2015. Existential Risks: Exploring a Robust Risk Reduction Strategy. *Science and Engineering Ethics*. 21:541. Jebari adds that "studies have shown that this is a roughly the number of people that colonized the Americas."

144 See Carrington, Damian. 2002. "Magic Number" for Space Pioneers Calculated. *New Scientist*. URL: https://www.newscientist.com/article/dn1936-magic-number-for-space-pioneers-calculated/.

145 Baum, Seth, David Denkenberger, and Jacob Haqq-Misra. 2015. Isolated Refuges for Surviving Global Catastrophes. *Futures*. 74: 45–56. Although see Turchin, Alexey, and Brian Patrick Green. 2017. Aquatic Refuges for Surviving a Global Catastrophe. *Futures*. In press.

146 Baum, Seth, David Denkenberger, and Jacob Haqq-Misra. 2015. Isolated refuges for surviving global catastrophes. *Futures*. 74: 45–56.

147 Beckstead, Nick. 2015. How much could refuges help us recover from a global catastrophe? *Futures*. 72: 36–44.

148 That is, given the "supreme emergency" that civilization would be facing. See Walzer, Michael. 1992. *Just and Unjust Wars: A Moral Argument with Historical Illustrations*. New York, NY: Basic Books. I would also recommend Coady, C.A.J. 2014. The Problem of Dirty Hands. *The Stanford Encyclopedia of Philosophy*. URL: https://plato.stanford.edu/archives/spr2014/entries/dirty-hands/. A somewhat related issue is what Karl Popper calls the "paradox of tolerance," which Popper describes as follows: "Unlimited tolerance must lead to the disappearance of tolerance. If we extend unlimited tolerance even to those who are intolerant, if we are not prepared to defend a tolerant society against the onslaught of the intolerant, then the tolerant will be destroyed, and tolerance with them." Popper, Karl. 1945. *The Open Society and Its Enemies, volume 1, The Spell of Plato*. UK: Routledge. In a WTD-cluttered world where a single individual could terminate the experiment of civilization on her or his own, it may *perhaps* become necessary to implement "zero tolerance" policies with respect to individuals who harbor a death wish for humanity or destruction wish for for civilization, such as those in endnote 396. Civilization cannot exist if tolerance extends to those who would like to demolish it.

149 More generally, we can borrow the term "catastrophe-catastrophe tradeoffs" from Cass Sunstein to describe a class of predicaments in which one opts for a lesser catastrophe L that, if pursued, would obviate a greater catastrophe G. Sunstein illustrates this idea as follows: "The war in Iraq was defended as a way of avoiding catastrophe . . . If Saddam Hussein possessed weapons of mass destruction (and there

was certainly some ex ante chance that it did), the risk of catastrophic attack could not be dismissed. But the war itself created serious risks, and thus created risk-catastrophe tradeoffs, perhaps calling for an assessment of probability: What was the likelihood of serious risks from war, and what was the likelihood of catastrophic harm from refusing to make war? Some people believe, in fact, that the war has given rise to catastrophic risks of its own, thus creating catastrophe-catastrophe tradeoffs." He adds that "efforts to control emissions of greenhouse gases could easily be analyzed in similar terms." See Sunstein, Cass. 2007. The Catastrophic Harm Precautionary Principle. *Issues in Legal Scholarship*. 6(3). Nonetheless, we should be extremely careful about what Pinker refers to as "a pernicious utilitarian calculus." He writes that, for example, "utopian ideologies invite genocide [in part because in] a utopia, everyone is happy forever, so its moral value is infinite. Most of us agree that it is ethically permissible to divert a runaway trolley that threatens to kill five people onto a side track where it would kill only one. But suppose it were a hundred million lives one could save by diverting the trolley, or a billion, or—projecting into the indefinite future—infinitely many. How many people would it be permissible to sacrifice to attain that infinite good? A few million can seem like a pretty good bargain." Pinker, Steven. 2011. *The Better Angels of Our Nature: Why Violence Has Declined*. New York, NY: Penguin Books.

150 Bostrom, Nick. 2002. Existential Risks: Analyzing Human Extinction Scenarios and Related Hazards. *Journal of Evolution and Technology*. 9(1).

151 Harris, Sam. 2005. *The End of Faith: Religion, Terror, and the Future of Reason*. New York, NY: WW Norton & Company. See endnote 77 in chapter 4 for an important disambiguation of the term "Islamist."

152 Bostrom, Nick. 2002. Existential Risks: Analyzing Human Extinction Scenarios and Related Hazards. *Journal of Evolution and Technology*. 9(1).

153 In more detail, Muehlhauser elaborates that , "Differential intellectual progress consists in prioritizing risk-*reducing* intellectual progress over risk-*increasing* intellectual progress. As applied to AI risks in particular, a plan of differential intellectual progress would recommend that our progress on the scientific, philosophical, and technological problems of AI *safety* outpace our progress on the problems of AI *capability* such that we develop *safe* superhuman AIs before we develop (arbitrary) superhuman AIs. Our first superhuman AI must be a safe superhuman AI, for we may not get a second chance." See Muehlhauser, Luke. 2013.

Facing the Intelligence Explosion. Berkeley, CA: Machine Intelligence Research Institute. As for the term "benevolent global hegemony," coined by neoconservatives but useful in the present context, see Beaumont, Peter. 2008. A Neocon by Any Other Name. *Guardian.* URL: https://www.theguardian.com/world/2008/apr/27/usa.

Chapter 7

1   Bell, Wendell. 1993. Why Should We Care about Future Generations? In H.F. Didsbury, Jr. (editor), *The Years Ahead: Perils, Progress, and Promises.* Bethesda, MD: The World Future Society.

2   Another issue not here discussed is the possibility that a number of small disasters produce a synergistic cascading cause-and-effect stream that ultimately leads to the collapse of civilization. For example, Bruce Tonn and Donald MacGregor "describe a scenario in which a series of global catastrophes lead to a rapid decrease in human population. Societies become Balkanized; the previous 'Haves' retreat to enclaves and destroy the technological infrastructures outside of the enclaves that could be used by the 'Have-nots' to threaten the enclaves. Over time, the enclaves themselves perish because inhabitants outlive their capacity for fertility and human cloning fails. Resources available to the 'Have-nots' continue to collapse. A series of natural disasters then befall the Earth and the remaining humans. Due to rapid cooling and the oxidation of suddenly exposed rock from falling sea levels, the level of oxygen in the atmosphere drops below the level needed to sustain human life. The last humans asphyxiate. This scenario posits a long chain of unlikely events, yet this series of events—in any order—could plausibly lead to a major decline in population." More generally speaking, it could be than a risk scenario A nontrivially boosts the probability of another scenario B happening, thereby resulting in double, triple, or *n*-tuple catastrophe cascades: a domino effect that terminates with the total destruction of civilization. Quoted in Tonn, Bruce, and Dorian Stiefel. 2012. The Race for Evolutionary Success. Sustainability. 4:1787-1805. See also Tonn, Bruce, and Donald MacGregor. 2009. Are We Doomed? *Futures.* 41(10): 706–714.

3   Bostrom, Nick. 2002. Existential Risks: Analyzing Human Extinction Scenarios and Related Hazards. *Journal of Evolution and Technology.* 9(1). More informally, Bostrom appears to estimate a roughly 17 percent chance of human extinction before 2100 on a whiteboard in the Future of Humanity Institute's offices. See https://www.flickr.com/photos/arenamontanus/14427926005/in/photolist-axC1R7-5LFwwU-7hQJqk-

9bt9At-pY8ypH-nYWTJV-hS8HiV-kDqhCb-cxqxN1-cxqrZJ-pBqeDj-odfdF2-4DqCXj-f3rfff-mPsvky-6qqYwu-cSuQJu-c4jqZ5-6Jaj5R-9V-Ygo7-jzAnZC-gtTN7P-uZ3z1K-vHKd3U-qqsUqQ-7cUu4M/.

4   Again, not including Leslie's estimate of a 30 percent chance of doom within the next 500 years.

5   Hawking, Stephen. 2016. This Is the Most Dangerous Time for Our Planet. *Guardian.* URL: https://www.theguardian.com/commentis-free/2016/dec/01/stephen-hawking-dangerous-time-planet-inequality. (A paragraph break was deleted.) Indeed, in terms of food production, it appears that *we will have to grow more food in the next 50 years than in all of human history*! See Potter, Ned. 2009. Can We Grow More Food in 50 Years Than in All of History? ABCNews. URL: http://abcnews.go.com/Technology/world-hunger-50-years-food-history/story?id=8736358.

6   Leslie, John. 1996. *The End of the World: The Science and Ethics of Human Extinction*. New York, NY: Routledge.

7   More precisely, imagine that a "referee" flips a fair coin to decide which bucket you are going to draw from and that you're unable to see the outcome. Thus, you don't know which bucket from which you are drawing. Thanks to Daniel Kokotajlo for this.

8   Bostrom, Nick, and Milan Ćirković. 2003. The Doomsday Argument and the Self-Indication Assumption: Reply to Olum. *The Philosophical Quarterly*. 53(210): 83–91.

9   Bostrom, Nick. 2002. *Anthropic Bias: Observation Selection Effects in Science and Philosophy*. New York, NY: Routledge. See also Bostrom, Nick. 2005. Self-Location and Observation Selection Theory: An Advanced Introduction. URL: http://www.anthropic-principle.com/pre-prints/self-location.html.

10   See Armstrong, Stuart. 2011. Anthropic Decision Theory for Self-Locating Beliefs. URL: https://pdfs.semanticscholar.org/fa3d/1d0b733e52940c85faca24019c79f32e90df.pdf.

11   The anthropics scholar and AI researcher Katja Grace was the first to point this out. See Grace, Katja. 2010. SIA Doomsday: The Filter Is Ahead. *Meteuphoric.* URL: https://meteuphoric.wordpress.com/2010/03/23/sia-doomsday-the-filter-is-ahead/; and Grace, Katja. 2010. Anthropic Reasoning in the Great Filter. BS honors thesis, Australian National University. URL: https://dl.dropboxusercontent.com/u/6355797/An-thropic%20Reasoning%20in%20the%20Great%20Filter.pdf.

12   Quoted from Hanson, Robin. 2010. *Very* Bad News. *Overcoming Bias*. URL: http://www.overcomingbias.com/2010/03/very-bad-news.html.

318 • Notes to Chapter 7

13 Specifically, they are theories about how one should update one's credences given the evidence. For an excellent discussion of these esoterica, see Kokotajlo, Daniel. 2017. The Lesser Evil: A Defense of Self-Sampling and Centered Conditionalization. MA Thesis, University of North Carolina at Chapel Hill.

14 For a putative refutation of the doomsday argument, see chapter 7 of Häggström, Olle. 2016. *Here Be Dragons: Science, Technology and the Future of Humanity*. Oxford: Oxford University Press. But see also Bostrom, Nick. 1999. The doomsday argument is alive and kicking. *Mind*. 108(431): 539–551.

15 Jacquet elaborates the idea as follows: "Built as it is from the Greek *anthropos*, the term *Anthropocene* implicates all of humanity. This could be a dangerous way of seeing ourselves. Studies of the placebo effect have long shown that if we believe an inert pill to be real, some will experience suggested benefits. But there is also the so-called nocebo effect, in which some will experience suggested negative side effects of a placebo. Naming the Anthropocene might work the same way, bringing to bear an alarming fait accompli: human destruction could be exacerbated if we believe that destruction is what we do—that environmental destruction is a byproduct of human nature. Call it the anthropocebo effect." Jacquet, Jennifer. 2016. Human Error: Survivor Guilt in the Anthropocene. *Lapham's Quarterly*. URL: http://www.laphamsquarterly.org/disaster/human-error. See also and Jacquet, Jennifer. 2013. The Anthropocebo Effect. Edge.org. URL: https://www.edge.org/response-detail/23701.

16 Romer, Paul. Conditional Optimism about Progress and Climate. Accessed February 5, 2017. URL: https://paulromer.net/conditional-optimism-about-progress-and-climate/.

17 Quoted in Torres, Phil. 2017. The Only Way to Prevent Another Nuclear Strike Is to Get Rid of All the Nukes. *Motherboard*. URL: https://motherboard.vice.com/en_us/article/doomsday-clock-interview-lawrence-krauss.

18 Some formatting changes were made to this quote. Romer, Paul. Conditional Optimism about Progress and Climate. Accessed February 5, 2017. URL: https://paulromer.net/conditional-optimism-about-progress-and-climate/.

19 As Bostrom puts it, "The point, however, is not to wallow in gloom and doom but simply to take a sober look at what could go wrong so we can create responsible strategies for improving our chances of survival." See Bostrom, Nick. 2002. Existential Risks: Analyzing Human Extinction Scenarios and Related Hazards. *Journal of Evolution and Technology*. 9(1).

Postscript

1   Or, as the technology critic Lewis Mumford put it, "I would die happy if I knew that on my tombstone could be written these words, 'This man was an absolute fool. None of the disastrous things that he reluctantly predicted ever came to pass!'" Quoted in Jensen, Derrick. 2016. *The Myth of Human Supremacy*. New York, NY: Seven Stories Press.

2   Kelly, Kevin. 2008. The Expansion of Ignorance. The Technium. URL: http://kk.org/thetechnium/the-expansion-o/.

3   See also Winner, Langdon. 1977. *Autonomous Technology: Technics-out-of-Control as a Theme in Political Thought*. Cambridge, MA: MIT Press.

4   Again, where "knowledge" means "that known by collective humanity." For more, see my paper (written under a now-defunct pen name) Verdoux, Philippe. 2011. Emerging Technologies and the Future of Philosophy. *Metaphilosophy*. 42(5): 682–707.

5   Quoted in Rosenberg, Jay. 2009. Wilfrid Sellars. *The Stanford Encyclopedia of Philosophy*. URL: https://plato.stanford.edu/archives/sum2011/entries/sellars/.

Index

About the Author

Phil Torres is the founding director of the X-Risks Institute. He has written about apocalyptic terrorism, emerging technologies, and existential risks for a wide range of media, both popular and academic, including the *Bulletin of the Atomic Scientists, Time, Skeptic, Free Inquiry, Motherboard, Journal of Evolution and Technology, Journal of Future Studies, Erkenntnis, Bioethics, Foresight,* and *Metaphilosophy.* His previous book was *The End: What Science and Religion Tell Us about the Apocalypse.* Follow him @xriskology.